CRUDE RUBBER

AND

COMPOUNDING INGREDIENTS

CRUDE RUBBER

AND

COMPOUNDING INGREDIENTS

A TEXT-BOOK OF

RUBBER MANUFACTURE

HENRY C. PEARSON

GYAN

Published by

Gyan Publishing House
5, Ansari Road
Daryaganj, New Delhi-110002
Phone: 011-47034999, 9811692060
E-mail: books@gyanbooks.com

Distribution Network
gyanbooks.com
India, USA, Canada, UK, Australia, France

ISBN: 978-81-212-9404-1 (PB)
First Published, 1899

2nd Impression 2023

Printed at: Gyan Press, Delhi.

CRUDE RUBBER AND COMPOUNDING INGREDIENTS

AUTHOR: HENRY C. PEARSON

CRUDE RUBBER

AND

COMPOUNDING INGREDIENTS

A TEXT-BOOK OF

RUBBER MANUFACTURE

BY HENRY C. PEARSON

Editor of *The India Rubber World*

NEW YORK AND LONDON

The India Rubber Publishing Company

1899

PREFACE.

In its beginning, this volume was simply a brief, conveniently arranged portfolio of facts concerning the various ingredients that enter into rubber compounds. It was designed for the author's use alone. Certain rubber manufacturers, however, learning of its existence, and desiring copies of it, urged strongly that the matter be published and given to the trade as a whole in book form. The superiority of such a collection over the most comprehensive book of compounds doubtless will be apparent to the expert manufacturer, for this reason. When a manufacturer buys a set of compounds—and most of them are purchasable—he invariably acquires them not so much for use as for suggestion and comparison. The descriptions, therefore, of a great majority of the ingredients used in all lines of rubber compounding, and scores with which he may be unfamiliar, will be so suggestive to the practical man that new sets of compounds will be secured, each partaking of the individuality of the expert, and bearing the impress of the line of work done in the factory to which he is attached, and wholly free from the taint of imitation or counterfeiting, which is the bane of the purchased secret. It is felt that another point of superiority over the mere compound book will be found in the fact that no private formulas are given, which might wound the feelings of the more conservative manufacturers.

The higher level of prices for crude rubber which has prevailed for some time past is drawing the attention of manufacturers to the pseudo gums—such, for instance, as Pontianak. A number of these are described in this volume, with the hope that, since their presence in the market depends largely upon the insistence with which rubber manufacturers demand them from importers or gatherers, many more may be made generally useful. Another line of endeavor that already has been fruitful of excellent result is in the production of rubber substitutes. It is hoped that the chapters upon this subject, covering some two hundred varieties, some excellent and others perhaps of little value, may prove of such service that, even if the ideal substitute does not materialize, the knowledge of the scope of the substitutes already known will be so increased that their intelligent use will be greatly amplified.

In the compilation of this book free use has been made of English, German, and French standard technical works as well as of technical journals, such as *The India Rubber World, The India-Rubber and Gutta-Percha Trades Journal,* the *Gummi-Zeitung, The Journal of the Society of Chemical Industry,* and others.

The author takes pleasure in acknowledging his indebtedness for helpful suggestions to skilled manufacturers and superintendents in both America and Europe, and to the following distinguished writers on rubber topics: P. G. W. Typke, F. C. S.; G. S. Jenman, Government Botanist and Superintendent of the Botanic Gardens, Demerara; William Thompson, F. R. S. E.; H. Grimshaw, F. C. S.; W. Lascelles-Scott, F. R. M. S., M. S. C. I.; Richard Gerner, M. E.; Dr. C. Purcell Taylor, Thomas Bolas, F. C. S., F. I. C.; Professor D. E. Hughes, F. R. S.; Messrs. Heinzerling and Pahl, Berlin; Granville H. Sharpe, F. C. S.; Carl Otto Weber, Ph. D.; A. Camille, J. H. Hart, Superintendent Botanic Gardens of Trinidad; Dr. D. Morris, M. A., C. M. G., Commissioner of the Imperial Agricultural Department for the West Indies; the late Dr. Eugene F. A. Obach, F. I. C., F. C. S., M. E. E.; Professor F. A. C. Perrine, D. Sc., and many others.

I also wish to express my appreciation of the valuable assistance given me on the chapters devoted to crude India-rubber and Gutta-percha, and the assistance in editing and revising other portions, by my associate editor on *The India Rubber World,* Mr. Hawthorne Hill.

Boston, June, 1899. HENRY C. PEARSON.

CONTENTS.

CHAPTER I.

GRADES OF CRUDE RUBBER, SOURCES OF SUPPLY, AND PHYSICAL CHARACTERISTICS.

CAOUTCHOUC or India-rubber is a product of a great variety of trees, vines, and shrubs, most of which grow in the torrid zone. Central America, South America, Africa, and India all furnish their quota, and while the gum that comes from these vast areas is all rubber, it differs widely in its characteristics, due in a measure to a variety of methods in gathering and coagulation, but more specifically in its chemical constituents. South America produces the best rubber in the world and the most of it. The Amazon valley, embracing hundreds of thousands of square miles of rubber forests in Brazil, Bolivia, and Peru, is the center of the industry, the product being exported from the city of Para, whence the name "Para rubber." Two or more species of the *Hevea* produce this rubber, the best known being the *Hevea Brasiliensis.* Peru also produces a rubber, lower in grade than Para, known as "Caucho." The *Castilloa elastica,* the rubber tree of Nicaragua and other Central American states, which is also found in Ecuador, Venezuela, Colombia, and Mexico, produces the rubber known as "Centrals." The Atlantic states of Brazil, south of Para, produce other rubber trees, from which come the grades known as "Mangabeira," "Pernambuco," and "Ceara." Africa comes next to South America in the amount of rubber produced, and in the interior of that continent, as in the Amazon country, there are great rubber forests as yet untouched. African rubber is inferior to that obtained from South America, but through improved processes in gathering and curing, the various sorts are delivered in much better condition year by year. African rubber is found on both the east and west coasts and throughout the great basin of the Congo river and also on the island of Madagascar.

The *Landolphia,* of which there are several species, is a giant vine or creeper from the milk of which most of the African rubbers come. Rubber from Lagos and from some other colonies in West Africa, however, is obtained from a tree known as the

Kickxia Africana. The East Indies to-day furnish but little rubber. The first rubber exported from that part of the world came from Assam. In time, however, Burma became a producer of a similar grade, known as "Rangoon" rubber. The principal source of rubber in that part of the world is a tree known as the *Ficus elastica.* The islands of Java and Borneo and also Penang and other states in the Malaysian peninsula produce a certain amount of rubber.

Seaports, trading posts from which the first shipment is made, the name of a colony or country, or descriptive terms, as "thimbles," "buttons," "strips"—all or any of these may serve for names of different grades of crude rubber. A complete market report would indicate that there are a great number of different qualities of rubber, many coming from the same source. This, however, is not wholly true. Take, for instance, the Para grades: years ago any rubber coming from Brazil was called Para rubber. Later it was divided into "fine," "medium," and "coarse." Then the rubber from the islands in the lower Amazon became known as "Islands rubber," while that coming from further up stream was known as "Upriver," and these, too, were divided into fine, medium, and coarse. Now a dozen or more local names are applied to rubber from different localities, tributary to the Para market. At the same time, most of these rubbers sell at the same figures, grade for grade, with the exception of coarse.

Something like this is true in the African rubber trade. For instance, a great number of local names are applied to the Congo rubber. The difference between "Equateur," "Kassai," and "Lopori" sorts may not be greater than between different lots from the same place. With a very few exceptions, the names which follow are those used commonly in the leading markets:

PARA RUBBER.

RUBBER is classified at Para and Manaos into three grades, designated by the Portugese words *fina, entrafina,* and *sernamby.* These same grades in the United States are known as "fine," "medium," and "coarse," while in England they are classified as "fine," "entrafine," and "negroheads," the latter being divided to provide for a subgrade, "scrappy negroheads." The production is about 60 per cent. of fine, 13 per cent. of medium, and 27 per cent. of coarse.

FINE PARA rubber comes in large bottles and, when cut, shows a surface closely marked with lines, corresponding to the number of layers of rubber milk added during the smoking process. These layers are easily separated and, when stretched, are very transparent. This rubber smells not unlike smoked bacon.

MEDIUM or ENTRAFINE resembles "fine," but is not so well cured, curds and globules of milk not perfectly smoked being found between the layers.

COARSE or SERNAMBY is made up of the residue, scraped daily from the collecting vessels, or from milk which has curdled before it could be smoked and made into "fine."

Besides this general classification of Para rubber, other names are in use, derived from the localities of origin.

ISLANDS rubber is that produced on the island of Marajo, some 17,500 square miles in extent, and other islands in its vicinity in the delta of the Amazon, together with that from other parts of the state of Para, except the Xingu, Tocantins, and Tapajos rivers, which might well be called lower Amazon grades. These islands in a recent year yielded over 12,000,000 pounds, or 63 per cent. of the total for the state of Para. The Islands "fine" and "medium" rubber is in the form of round or flat bottles, while the "coarse" or "sernamby" is in scraps massed into balls and round cakes, which gives the name "negroheads" to this grade in the English market.

CAVIANA rubber, named from the island that produces it, is the highest grade of Islands, and is to-day marketed as a distinct sort. It has a smooth close grain, and is much in demand for fine work.

CAMETA rubber is so called from the port of that name, on the Tocantins river. It is noted for the superior quality of its "sernamby" grade, the "fine" being the same as from the islands, but rarely seen. This rubber comes in the form of little cups pressed into large "negroheads."

UPRIVER rubber includes the product of the country bordering the Amazon and its tributaries above Para, and that which comes from Peru and Bolivia through the large streams rising in those countries—such rivers are the Purus, Jurua, Javary, and

Madeira. This rubber for the most part is derived from the *Hevea discolor* and comes to market in biscuits varying greatly in size and shape, a full average biscuit weighing about thirty pounds. The rubber tree on the islands is more frequently the *Hevea Brasiliensis*, but it is a mooted question whether the difference in the trees accounts for the difference in quality between Upriver and Islands rubber. "Upriver" rubber is marketed also under such local names as "Manaos," "Madeira," "Bolivian," etc.

ITAITUBA rubber comes from the port of that name, at the head of steam navigation on the Tapajos river, which enters the Amazon at Santarem. Rubber from this river is distinguished for the rather gutty quality of the "fine" and "medium," and its stringy, dirty "sernamby."

XINGU rubber from the Xingu river, is noted for the specially good cure of the "fine."

MANAOS rubber is named from the city which is the capital of Amazonas, 1,200 miles up the Amazon river, and the center of the rubber trade of an immense district. Upriver rubber exported direct to foreign markets from this port is sometimes designated as "Manaos rubber."

MADEIRA rubber, named from a great river which joins the Amazon below Manaos, is of excellent quality and produced in large quantities. It has a finer and closer grain than any other upriver rubber except the Bolivian.

BOLIVIAN rubber is floated down the Beni and other rivers in Bolivia to the Madeira, and thence to the Amazon. It meets innumerable detentions from cataracts in the upper Madeira, on account of which it becomes somewhat dried before reaching market. It has the further advantage of being cured by a better class of labor than is common in Brazil, of having a tougher fiber and of being cleaner than most upriver rubber, for which reasons it brings higher prices than any other.

MOLLENDO rubber comes from southern Bolivia, being transported by steamers across Lake Titicaca and by rail to Mollendo, a Peruvian port on the Pacific, and thence principally to England. It is prepared in biscuits and sheets and is marketed at a price between upriver and islands.

ANGOSTURA rubber comes down the Orinoco in Venezuela,

from Cuidad Bolivar, which town formerly was known as Angostura. It is of the same grades as the Para sorts. Some of the same class of rubber finds its way into Brazil, at Manaos, where its identity is lost.

ORINOCO rubber is the same as "Angostura."

MATTO GROSSO rubber is from the state of that name in the southwest of Brazil, and reaches the market partly through tributaries of the Amazon and partly through the Parana, which discharges into the river Plate. It comes in "fine," "medium," and "coarse," but principally the latter, little of it reaching the market at present.

CAUCHO is a distinct sort of rubber, inferior to that from Para, collected along the Peruvian rivers tributary to the Amazon and particularly along the Javary. It is not cured by smoking, but by the admixture with the milk of lime, potash, or soap. The physical characteristics of Caucho in the main are the same as in the Central American rubbers. It is known also as "Peruvian rubber" or "Peruvian caucho." It is exported from Iquitos, Manaos, and Para, and included in the general total of rubber exports from the Amazon country. It comes to market in three forms—Ball, Strip, and Sheet (or Slabs)—ranging in value in the order named.

CENTRAL RUBBERS.

CENTRAL AMERICAN rubber, or "Centrals," includes that which is produced in all the states north of the Amazon valley, up to and including southern Mexico. It forms a distinctive class, being the product of a tree not found elsewhere. The consumption of Centrals in the United States was larger once than of Para rubber, but the yield has declined gradually to small proportions. This rubber is in good demand for certain uses, ranking in price below coarse Para. It has not the toughness or strength of fine Para, and possesses less elasticity. Centrals are classed usually as "sheet" and "scrap," besides which the terms "strip," "slab," "ball," and "sausage" are used. Greytown being a common shipping-point for Centrals, there is much confusion, one sort often getting substituted for another. Most of the yield of Costa Rica is exported through Nicaragua. The treatment of Centrals generally consists in heating the sap and stirring in a strong concoction of the mik of bindweed, the product being "sheet" rubber. The rubber drippings which adhere to the bark of the tapped trees are peeled off when dry and called "scrap." The trade names below apply to the locality of origin, rather than indicating distinctions in quality.

NICARAGUA rubber includes more than the product of that republic. The real Nicaragua rubber is drier as a rule than other grades of Centrals. Nicaragua sheet comes to market in a less clean condition than formerly, and the scrap now brings a better price.

GREYTOWN SCRAP is the best grade of Nicaragua rubber.

GUATEMALA rubber is inferior and unequal in quality. The best is whitish in color, and the lower grades black with a tarry appearance. It is said to be sometimes adulterated with cheap molasses. In curing, the rubber-gatherers pour the sap upon mats to dry, afterwards pulling off the product in sheets, pressing them together for shipment.

GUAYAQUIL STRIP, from Ecuador, is imported in two grades —good and ordinary. Like the Guatemala rubber, the best has a whitish appearance. The inferior sort is porous and filled with a fetid black liquid, which carries an almost indelible stain.

ESMERALDA rubber, which also comes from Ecuador, is classed as a Strip and Sausage, the two grades coming to market in about equal quantities.

COLOMBIAN is a pressed strip rubber, dark in color, sometimes showing white when cut. It is graded "No. 1" and "No. 2." Some of the rubber from Colombia bears local designations, besides varying in quality. These include:

Cartagena, a strip rubber, dark and tough, graded "No. 1" and "No. 2," selling at less than "Colombian." It comes also in thin sheets, rough or "chewed" in appearance, and tarry or sticky. The production has decreased very much of late.

Panama rubber, like that from Nicaragua, embraces a wide range of quality. The Pacific mail steamers bring together at Panama rubber from numerous ports, and confusion of grades is a result. What is marketed as "Panama" comes in "sheet" and "strip."

Tumaco comes in "sheet," "slab," and "scrap," from the Pacific coast of Colombia. Very little of it is received.

MEXICAN rubber is of fair quality, but is received in constantly decreasing quantities. The grades, listed in the order of their selling value, are Ball, Strip (or Scrap), and Slab.

Tuxpam strip comes from the Mexican port of that name. Very little of it is received, and that not of uniform quality.

HONDURAS STRIP is of a quality similar to the Mexican, but is little produced.

WEST INDIAN rubber has a good reputation for quality. It is not produced on the islands, but comes from Venezuela and Central America, and is simply a general trade name used in England.

The grades which follow, though not entitled geographically to be included as "Centrals," are in fact so classed, on account of their quality.

MANGABEIRA rubber is so called from the local name of the tree producing it, in the Atlantic states of Brazil, south of Para. It is an alum-cured rubber and comes in sheets, which resemble slices of liver and are of a tawny red color. The thin sheet sells for more than the thick, as it is dryer and better cured. Occasionally it comes in the form of balls. It is exported from Pernambuco, Bahia, Natal, and other points on the coast.

PERNAMBUCO is another name for Mangabeira rubber, derived from the principal state and port from which it is shipped.

CEARA rubber comes from a tree particularly abundant in the Brazilian state of Ceara and is marketed principally in England. The sap exudes from the tree and coagulates in the form of "tears" which are gathered in scraps and balls. There are three grades, the lowest of which is dirty and difficult to use. Ceara rubber is deficient in elasticity and is hard to vulcanize. It is very dry and free from stickiness.

AFRICAN RUBBER.

AFRICAN rubbers, though comparatively late in becoming known, are produced now in quantities second only to the supply from the Amazon. As a class they are more adhesive and less elastic than Para rubbers, ranking with or below Para negroheads. They often contain a liberal percentage of impurities, and for a long time their disagreeable odor and intractable nature hindered their introduction. But advancing prices for Para grades and fear of their coming scarcity led manufacturers to experiment with African rubbers, until many uses were found for them. The result has been a temporary check in the upward tendency in price of the Para grades, although there are many purposes for which Africans never have been considered as competing with them. At the same time, the possibilities in the way of utilizing African sorts have not been exhausted, each year bringing out new uses.

The African rubbers are obtained from giant creepers, of which there are a dozen species on the continent and in the island of Madagascar, and also from several trees, the most important one of which abounds on the Gold Coast, in Lagos, and some other West Coast colonies. The adulteration of African rubbers is not uncommon, being due to the dishonesty, not only of the native gatherers, but doubtless also of some foreign traders on the coasts. But in several of the English and Belgian colonies stringent laws have been passed to prevent such adulterations. On the Gold Coast the lumps of rubber brought to market by the natives were formerly cut into strips or buttons by machinery, before being exported. To-day this work is done in England, the rubber then being known as "Liverpool pressed." It has been urged by some importers of Lagos rubber that wilful adulteration by the natives is rare. Rubber has been worked in Lagos for only about four years, so that many of the workers there are yet inexperienced and lacking in skill. Even in the Gold Coast Colony, where the industry began ten years earlier, a certain percentage of the rubber is spoiled in gathering.

The milk of the *Landolphia* vines, the chief rubber producers of Africa, coagulates on exposure to the air, though in some localities use is made of various astringents, boiling in water, and other methods to assist in preparing rubber. Even where these methods are used, a residue of the rubber sap is left to dry on the bark and in the earth, and is gathered in strings or scraps. The only treatment in some other places is the smearing of the sap upon the bare bodies of the natives, where it dries speedily in the sun, and is easily peeled off.

BALL is the classification of a large share of the African rubbers, which comes in every size from three or four inches in diameter down to half an inch or less. "Small ball" of the several kinds differs from the "large ball" in size, and is also dryer and affords a smaller degree of shrinkage.

THIMBLES.—The natives, after gathering this rubber, cut it into cubes, about an inch square or less. Thimbles contain bark and sand, but very little moisture.

NUTS.—Rubber thimbles from Ambriz are quoted sometimes in European markets as "Ambriz nuts."

LUMP rubber comes in large pieces, varying in size and of irregular shapes. When packed in casks the pieces often become massed together in transit. It is from the best of the lump rubber that the most desirable buttons and strips are made.

FLAKE comes in lumps, livers, and soft irregular masses, and is valuable in the factory chiefly for frictions and for softening compounds.

PASTE is the same as "Flake." The Accra flake and Niger paste, which are the same in quality, are at the foot of the list, in respect to prices, the Niger being the cleaner.

STRIPS are lump rubber that is sliced and pressed by machinery before it is offered to the trade.

BUTTONS is a name applied to rubber similarly treated as in making strips, except that it is cut into small pieces, whereas strips have been marketed in every length up to ten feet.

BISCUITS is another name for "Buttons."

OYSTERS is another name for "Buttons" or "Biscuits."

TONGUES.—Some rubber formerly came to market in long, narrow, tongue-shaped pieces. The same grades are now more frequently seen in the shape of large balls.

NIGGERS are of various sorts and from different sources. These rubbers are ball-like in some cases, having the appearance of masses of stringy rubber pressed together between the hands and wound into compact masses.

TWIST rubber is not unlike "Niggers" in quality, but shows less shrinkage and differs in preparation and appearance. The string or strip-like pieces are wrapped about each other in order to give a twisted look to the balls.

The list of rubber grades which follows is based upon a geographical arrangement, beginning with the upper west coast of Africa:

GAMBIA.

Gambia Niggers (No. 1, No. 2, No. 3).—These are classified according to cleanliness, No. 1 and No. 2 being fairly clean, and No. 3 containing considerable soil.

Bathurst.—Same as Gambia.

SIERRA LEONE.

Sierra Leone Twists (No. 1, No. 2, and rejections).—This is white and amber in color, of low shrinkage, and has bark and grit in it, but little moisture.

Niggers (No. 1, No. 2, No. 3) are quite moist. No. 2 and No. 3 contain considerable soil.

Cake.—Fairly clean, but wet. It is both red and white, the former bringing the better price.

Manoh Twists.—This comes in the shape of tightly wound cords of rubber and works soft. In color it is black or white, the black being the best.

LIBERIA.

Liberian.—This is graded as Lump, Hard Flake, and Soft. It cuts yellow, is very wet, and is often a soft pasty rubber.

ASSINEE.

What is known as Assinee is graded as follows: *Assinee-Silky, Grand Bassam, Attoaboa, Lahou, Bayin, Half Jack.* It is like Old Calabar, only it comes in chunks three inches square, is wet, and cuts yellow. These names are chiefly used in the English market.

GOLD COAST COLONY.

Gold Coast.—This is chiefly lump from which Strips and Buttons are made. There are also Biscuits and Niggers (hard and soft). The Flake is wet and has a bad smell, but otherwise is quite clean.

Accra.—The Accra lump furnishes Strips and Buttons and is graded "prime," "seconds," and "thirds." The lower grades are Flake and Paste.

Cape Coast.—This is another lump from which Strips and Buttons are manufactured and has for lower grades Flake and Soft.

Salt Pond.—This Lump is also used in Strips and Buttons, the lowest grade being Flake.

Addah Niggers (graded as No. 1 and No. 2) is very similar to Sierra Leone, but generally in smaller balls. It is not an Accra rubber, nor are *Quittah Niggers* or *Axim*. As a matter of fact, the grades from these different ports differ little if any, and are sold most frequently under the head of "Accra" rubber, from the name of the principal town in the colony.

TOGOLAND.

Lomi (or *Lome*) *Ball.*—The best grade of this is a clean, firm rubber and is fairly dry. The lower grades are rarely seen.

LAGOS.

Lagos.—This Lump is also turned into Buttons and Strips, while soft inferior lumps are sold without manufacturing, as low grades. It is very easily distinguished from Accra by its odor.

NIGER RIVER PROTECTORATE.

Niger.—The chief grade is Paste, which has an acid smell and is a low grade pasty rubber, wet but clean.

Old Calabar.—It is graded as Blue, Lump, and Niggers and is very bad smelling. The best lump is undoubtedly used for strips and buttons.

Benin Ball.—Is generally dirty and has a rotten, woody smell.

CAMEROONS (OR KAMERUN).

Cameroons.—The Ball is graded as large, mixed, and small; the Clusters, which contain some fifty balls, as No. 1 and No. 2; and the Knuckly ball, which is a small dry ball. This rubber has a fairly strong smell.

Batanga Ball ("B," "E").—Same as Cameroons, Batanga being the name of a river and country in the Cameroons.

FRENCH CONGO.

French Congo rubber is very similar to Cameroon, but the balls are larger.

Gaboon is the best known flake and has for additional grades: Lump, Large "O" Ball, and Small "O" Ball. The Flake is free from dirt and is soft.

Mayumba is both Ball and Flake. Another grade known as Mixed is a combination of the two and is sold as second quality.

Loango.—Ball.

These are names of rubber stations on the coast. The natives boil rubber milk, adding the juices of vines, and, while the rubber is hardening, wind it into balls, weighing from one-fifth pound to three pounds. The best rubber is not boiled, the milk drying on the wrists of the natives, as they tap the rubber vines. At the coast the balls are cut, to detect any cheating, and washed and packed in casks for export.

CONGO FREE STATE.

Congo rubber comes in the shape of Buttons, Balls (No. 1 and No. 2), Red Thimbles, and Black Thimbles. The Ball is similar to Cameroons, but tougher. The Dutch Congo Ball is the same as the Congo Ball, but is known as the best grade of that rubber. There is also the Congo (Kassai), Black Twist (graded as fine, mixed, and secondary), and Red Twist. The Strips are among the toughest of African rubbers and are dry, with a woody smell.

From the Lower Congo comes also the Luvituku, which is a Red Ball rubber, and from the Upper Congo, the following:

Upper Congo.—Ball, Red Ball, Twists, and Strips, all of which is good tough rubber.

Uelle.—Strips, usually heated and fermented and bad smelling; Cakes, wet, but clean.

Sankuru.—Ball, very similar to Congo Ball.

Lake Leopold.—Graded as Sausage and Ball. It does not differ from the foregoing enough to warrant special description.

Equateur.—In the form of balls (small and mixed). It is dark, dry, and clean, but contains some fermented rubber, which smells badly.

Lopori.—Graded as Ball (large and small), Strips, and Cakes. Some of the balls are fine and clean, while others contain fermented milk. Lopori also comes as Sausage.

Bangui.—Comes in the form of strips and is a firm, tough rubber.

Bussira.—Ball; a trifle softer than Lopori, but usually of excellent quality and dry. In use it develops a strong smell.

Aruwimi.—Ball. This usually comes as large, firm balls, but on cutting them open much of the interior is found fermented.

Mongalla.—In this the Ball is similar to Upper Congo Red Ball. It also comes in Strips, and is a good rubber.

Bumba.—Ball; *Buki*—Ball; *Tava* and *Kwilu* are all good Upper Congo grades that are not distinctive enough to dwell upon.

Wamba.—This is a grade of Thimbles and is a good black rubber, with only ordinary shrinkage.

ANGOLA.

Benguela.—Graded as Sausage and Niggers. Of the latter, No. 1 is clean and tough, and No. 2 contains a large percentage of red leaf.

Loanda.—In this the grades, which are Sausage and Niggers, are similar to Benguela, but not so dry. There are also Twists (red and black).

Ambriz.—Chiefly Thimbles or Nuts; both are poor grades.

EAST AFRICA.

Mozambique rubber is that coming from the port of Mozambique, from other ports in the same colony, and perhaps from still other East African ports. It possesses some properties in common with the Madagascar rubbers. The rate of shrinkage is less than in most African sorts, and good prices are obtained. In the Liverpool market, which is the best for Mozambique grades, quotations are made for Orange Ball, Ball No. 1, Ball No. 2, Ball No. 3, Liver, Sausage, Root, Sticks or spindles, Sticks removed, Unripe.

The Orange Ball (resembling an orange in size and shape) is the choicest rubber. Other grades of Mozambique Ball are distinguished further as "white" and "red," the latter being inferior. Its reddish color is due to the fine bark mixed with it. The Unripe contains more bark than rubber, and is not thoroughly cured.

Sticks or spindles consist of spindle-shaped pieces made of slender strings of rubber wound around a bit of wood. Liver (or cakes) is in smooth pieces of irregular size.

Lumu Ball, Liver, Sausage, and Root come from the Mozambique port of this name. They are not rubbers of a distinctive sort.

MADAGASCAR.

Madagascar rubber ranks higher in price than most other African sorts. Considering the greater loss sustained in washing, it costs nearly as much at times as fine Para. It is a favorite with manufacturers of hard rubber, on account of the fine lustrous polish which it assumes under the buffing-wheel. The principal classification is between "pinky" and "black."

Pinky comes in round balls, weighing 1½ to 4 pounds, black on the outside from exposure to the air, but having a pinkish-white look when cut.

Black, also in small balls, when cut shows a dark color, and is more or less sandy and dirty.

Tamatave being the principal seaport, its name is liable to be applied to any grades shipped from there. But what is described as "Prime pinky Tamatave" is the best rubber produced in Madagascar.

Majunga rubber, from the west coast town of that name, is a dark rubber of special excellence, ranking next to "pinky" in price.

Niggers (or negroheads) are designated as "East coast" and "West coast," and also as "Red ball," and "Gristly." They generally contain sand and dirt.

Brown cure (or brown slab) is a still lower grade.

Unripe is the lowest. This term is applied to balls containing bark in the center.

Madagascar rubber is cured (1) by the use of salt water, in which case the water is never wholly expelled, leading to a heavy rate of shrinkage, and (2) by artificial heat. The island is rich in rubber forests, but the exports are restricted by the wasteful methods of the natives, which exhaust the trees and vines, particularly near the coast.

EAST INDIAN.

ASSAM rubber is strong and of firm texture. It is fairly elastic, though often less so on account of carelessness in gathering and the introduction of impurities. There are four grades usually (No. 1 to No. 4), of which the lower ones are extremely dirty and contain soft rubber. The better grades when cut have a glossy, marbleized appearance, somewhat pinkish in color. Assam rubber is marketed in small balls, made by winding up strings of rubber dried on the trees, and also in oblong slabs of irregular size, wrapped in plaited straw. The output has declined for several years, the attempts at the cultivation of new trees in Assam having been without practical results. Meanwhile the same species has been found in Burma, where the production of rubber has grown at an equal rate with the falling off in Assam.

RANGOON rubber is the product of Burma, exported through the port of Rangoon, and differs so little from Assam rubber as to require no separate description. Four grades are marketed, at practically the same prices as for Assam rubber.

JAVA rubber, from the island of this name, is dark and glossy, of a deeper tint than the Assam sorts, with occasional red streaks. Otherwise, its history and characteristics are nearly identical with those of Assam rubber. Three grades are recognized. The milk dries on the surface of the trees, on exposure to the air, and the shrinkage of the better grades is slight.

PENANG rubber (from one of the states in the Malaysian peninsula, including the island of Penang) is also very similar to that from Assam. There are three or four grades, at slightly lower prices than the Assam sorts bring.

BORNEO rubber ranks below the other Asiatic sorts, being lower in price, with a higher rate of shrinkage. It is of a whitish color, changing with age to a dull pink or red. It comes to market shaped like pieces of liver, and is soft, porous, or spongy. The pores are filled with salt water or whey, for the reason that salt is used to coagulate the rubber, and the water evaporating leaves a saline incrustation in the cells. There are three grades, the first of which is a good rubber, while the lowest, when cut, is almost as soft as putty, and is worth little.

GUTTA-SUSU is a local name applied in Borneo to what is known in the markets as "Borneo No. 3."

CEYLON SCRAP is the product of a few small plantations in Ceylon of the South American tree known as "Ceara rubber."

CHAPTER II.

SOME LITTLE KNOWN RUBBERS AND BASTARD OR PSEUDO GUMS.

For the last fifteen or twenty years reports have come in from all over the tropical world regarding the discovery of gums, some of which were similar to India-rubber, while others were more like Gutta-percha. In a few instances these gums have appeared on the market in due time under various names and have been useful. This is not the rule, however, and it is due to a variety of reasons. The first, perhaps, is the scientific attitude of those who primarily examine the samples received at the great centers of civilization. Unless gums are of high grade, and bear promise of being nearly as valuable as a good grade of India-rubber or Gutta-percha, they are usually pronounced as worthless, or nearly so. These same experts, it is well to remember, condemned reclaimed rubber and substitutes, which may lead the manufacturer to suspect that his wants are not always appreciated by the learned. It is possible, of course, that the scientists and experts are right, and that it would have been better had reclaimed rubber or substitutes never been known. Nevertheless, rubber manufacturers are ever in the market for them, and would welcome many of the pseudo gums and find large uses for them, if once they were within reach.

Aside from the scientific attitude is the indifferent attitude of the gatherers in their native wilds, and of the importers who see little profit in such cheap gums, and of the manufacturers themselves, who wait until a neighbor has tried something new before venturing to experiment.

It is only sufficient to recall what is needed in rubber compounding to see how many of these gums could be made valuable. For example, sometimes simple stickiness is called for; in another case only insulating qualities and stickiness; in still another, waterproofing qualities and stickiness; and it is well to add here, that where only one valuable quality exists in a gum others can often be supplied. As a matter of fact, in the present state of compounding and manipulation, the presence of resins is not heeded, short life can be overcome, and intractability can be done away with.

A few years ago a leading American rubber manufacturer attempted to secure from Mexico a quantity of the bark from a small tree which was believed to yield rubber, with a view to extracting the gum, by the boiling process. His agent, not understanding the instructions given, had enough of the shrubs cut off at the ground to make a steamer load, and shipped them entire—wood and all. A liberal yield was obtained of a gum equal in quality to a good grade of Centrals. The undertaking did not prove profitable enough, however, to cause it to be repeated. But without doubt it would pay to engage in the extraction of rubber from this shrub in the district where it abounds. More recently the writer has received a sample of gum, worth perhaps 35 cents per pound at present prices, which was the product of another Mexican shrub, said to be found in great quantities, and needing only harvesting and pressing to produce a valuable rubber.

It is with the hope that some of the gums mentioned in the following pages may be brought before the rubber manufacturers the world over, that space has been given to them.

SOME LITTLE KNOWN RUBBERS.

JEVE RUBBER.—Known only by hearsay. Probably the product of the *Siphocampylos Jamesonianus,* found in the valley of the Mayo, in Colombia, and also in Ecuador, and described by Humboldt.

COW TREE RUBBER.—The cow tree is very plentiful in tropical South America and yields a milk commonly used for food. This milk contains considerable caoutchouc, which is about 30 per cent. resin. Botanically it is known as the *Brosimum galactodendron.* [Dr. D. Morris, "Cantor Lectures," 1898.]

BAKA GUM.—Found in the Fiji archipelago. Comes from *Ficus obligua* (Foret). Used by natives for birdlime. Sap very abundant. Gum little known. Samples sent to England were reported upon as being suitable for mixing. As prices are to-day would be worth about 50 cents a pound. [Kew Annual Report, 1877.]

CUMAI RUBBER.—From the milk of a tree found on the Rio Negro and Uaupes, in Brazil. None comes to market. This milk is used by the natives for waterproofing purposes.

[Dr. D. Morris, "Cantor Lectures," 1898.]

MUSA RUBBER.—A gum expressed from the peel and leaves of the banana and pisang plants. No gum yet on the market. Process patented in England by Otto Zurcher, of Kingston, Jamaica. Also called "Banana Rubber."

MANDARNVA RUBBER.—A low grade of South American gum, somewhat like Ceara rubber. Little known. Is said to grow on the dry arid uplands of the interior. Is one of a number of gums that bear the native names, Cauchin, Pau, and Massaranduba.

[Revue Coloniale (Paris)—"Report on the State of Sao Paulo."]

ABBA RUBBER.—This is an African rubber, from Lagos. It probably is the product of the *Ficus Vogelii*. It is low grade rubber and cures soft and short. There is a large percentage of resin in the milk. [Dr. D. Morris, "Cantor Lectures," 1898.]

MANGA-ICE RUBBER.—Argentine republic. It is very abundant. Produces good rubber.

[E. L. Baker, Consul at Buenos Ayres, U. S. Consular Reports, 1892.]

MABOA GUM.—Said to be produced from a species of *Ficus* in Santiago de Cuba.

[Consul Reimer, United States Consular Reports, 1892.]

DURANGO RUBBER.—Said to be produced from a plant of the genus *Cynanchum,* belonging to the natural order *Asclepiadeae,* found in the Mexican state of Durango. A specimen was exhibited at the Philadelphia Centennial Exhibition in 1876. Probably identical with a rubber of which a sample was sent to the writer from Mexico in 1896. Very black, sticky, and full of vegetable matter. Would rank with a fair Accra flake.

[Henry H. Rusby, M. D.]

BRAZILIAN BIRDLIME.—The sap of the *Artocarpus incisa* is used by the Brazilians for birdlime and glue. When coagulated and dried the gum is white and somewhat similar to Gutta-percha. At ordinary temperatures it is hard and brittle, but with a little heat becomes plastic, and at the temperature of boiling water is soft and very sticky. It is soluble in bisulphide of carbon, and insoluble in alcohol and water. A similar gum of a chocolate brown color comes from the *Urostigma Gamelleira.*

[R. H. Biffen, Botanical Laboratory, Cambridge.]

BEIRA RUBBER.—Another name for stick rubber, gathered on the east coast of Africa, and shipped from Beira.

ROOT RUBBER.—A rubber obtained from the roots of a semi-herbaceous plant known as the *Carpodinus sanceolatus.* Very abundant in the open grassy country of the Congo Free State.

[Dr. D. Morris, "Cantor Lectures," 1898.]

AMAZONIAN RESIN RUBBERS.—The valley of the Amazon contains many trees and plants that are caoutchouc producers, but which are generally neglected, as the gatherers are seeking the more valuable *Hevea.* Among these are mentioned the trees known under the native names of Amapa, Sucuba, Surva, Tamanguiro, Molango, etc. All of these show a marked percentage of resin in the milk.

[Torres.]

BASTARD OR PSEUDO GUMS.

BALATA is the gum of the "bully" or "bullet" tree, found in British and Dutch Guiana, and in Venezuela. The Venezuelan product is known as "block" Balata; that from the Guianas as "sheet." Balata also differs in color, the white being considered better than the reddish. In character this gum occupies a position between India-rubber and Gutta-percha, combining in a degree the elasticity of one with the ductility of the other, and freely softening and becoming plastic and easily molded in hot water. The milk, diluted with water, is said to be drunk by the natives as a substitute for cow's milk. Balata is dried ordinarily by evaporation. A more rapid coagulation is effected by the use of spirits of wine. Alum and sulphate of aluminum are sometimes used to coagulate, but are not very satisfactory. The gum is sometimes mixed during the gathering with the milk that produces gum known as Touch-pong and Barta-Balli. Balata shrinks in washing from 25 to 50 per cent. It is used principally in the manufacture of belting and for insulation work. It has been utilized also for golf balls and as a substitute for India-rubber in dress shields.

PONTIANAK is a cheap inelastic gum imported from a town of the same name in Borneo. "Jelutong" is the import name in the United States, besides which the names "Fluvia" and "Gambria" have also been applied to it. The gum is used for a friction and filler. It is whitish in color, looking something like marsh-mallow candy, smells strongly of petroleum, and oxidizes readily on exposure to the air. It is believed to be the product of the tree known as the *Dyera costula.*

[Consul R. Wildman, United States Consular Reports, 1892.]

Tuno is a trade name of uncertain origin applied to a gum gathered principally in Nicaragua and Honduras. It is the product of what has been called the "sterile rubber tree" and also the "male rubber tree" of Nicaragua. The milk is coagulated with the aid of heat. The gum is but slightly elastic, is very sticky when heated, and is cheap. It is used as a friction gum, and is also mixed with Balata in the manufacture of belting. Sometimes is is sold under the name "Seiba gum," its identity being lost by ingenious massing and manipulation under water. Nicaragua rubber adulterated with "Tuno" in coagulation soon hardens and loses its elasticity. Also spelled "Toonu" and "Tunu."

Almeidina.—This comes from West Africa, particularly from the Cameroons and Angola, and has been found in the Solomon Islands. It is obtained from the tuber-like roots of a tree or shrub, and comes to market in small and sulphur-colored nodules, resembling potatoes, for which reason it has been called "potato gum." When broken open these balls look like putty, and although quite brittle when cold, the gum easily softens in warm water and may be drawn out in threads, which are possessed of some elasticity. It is completely melted at 240° F., and remains rather sticky after melting. It almost completely dissolves in cold benzine; in fact, nearly all of the solvents ordinarily used in rubber manufacture dissolve it. It mixes and dissolves with rubber in almost any proportion and up to 25 per cent. at least. Not only does it not injure the rubber, but is said to be beneficial to it. In working on the mill a pungent vapor arises from the mass, which, however, has no poisonous effect. In using this gum, a little caustic soda sometimes is added to the water when it is being washed; other manufacturers add tannic acid. Animal or vegetable fixed oils do not dissolve Almeidina, and, therefore when mixed with it are apt to rot it. Mixed with Gutta-percha this gum is practically indestructible. The name "Almeidina" is that of the first important shipper of the gum; in England the spelling "Almadina" has come into use. The gum is known also as "Euphorbia gum." [Thomas Christy and W. Lascelles-Scott.]

Gum Chicle.—A gummy resinous substance found around the seeds of the *Achras sapota*, a tree growing abundantly in the warm damp regions of Mexico and also in portions of Central

America. Chicle should be of a whitish color, odorous, and free from impurities, but often is adulterated with an inferior pink or reddish soil. It is solid and brittle at ordinary temperatures, but becomes plastic when placed in hot water. It is quite soft at 49° C. (120° F.). It is used chiefly in the United States in the manufacture of chewing gums, and to a small extent in England for adhesive plasters. It has been used for modeling purposes and for mixture with India-rubber for insulation work.

CATIVO GUM.—This comes from the sap of the mangrove called "Cativo" in the United States of Colombia. The gum is fluid at 130° F., and if the temperature is raised to 212° F. it is easily filtered and impurities removed, and a somewhat objectionable smell greatly lessened. The gum is then of a clear reddish brown color. It mixes easily with rubber and is said to produce a very tough compound. [Spon's Encyclopedia.]

TOUCHPONG GUM.—This is without doubt a rubber gum, entirely distinct from Balata. The rubber dries in strips on the trees, and what little of it comes to market has not been recognized as a distinct sort. Samples sent to England, however, have been favorably reported on. It is found throughout the Guianas. Probably from *Sapium biglandulosum*. Spelled "Touchpong" by Jenman; "Touchpong" by Morris; "Pouckpong" by Dr. Hugo Miller. [Dr. D. Morris, "Cantor Lectures," 1898.]

ABYSSINIAN GUTTA.—An adhesive acid gum of an earthy brown color, similar to common gutta in external appearance. Softens in water, but keeps a very great elasticity. On drying it remains exceedingly adhesive, therefore could not be used in place of Gutta-percha, but with proper treatment would undoubtedly make an excellent friction gum.

[Supplied by Mr. Thomas Christy.]

YELLOW GUTTA.—This comes from the Sunda Isles, from the genus *Payena*. It is practically a compound of India-rubber with two resins. One of these is crystalizable and the other is pitchy. If the raw material is treated with boiling alcohol the resins are taken off and the remaining product appears to be good India-rubber. [Edouard Heckel and Fr. Schlagdenhauffen, 1888.]

GUTTA GREK.—A gum that comes from Palembang, in Straits Settlements. It appears very much like India-rubber, but is permanently softened and destroyed by heat sufficient to melt it. It smells like Gutta-percha rather than India-rubber.

[T. Bolas in Colonial and Indian Exhibition Reports, 1887.]

GUTTA BASSIA.—Found between Upper Senegal and the Nile. Has the appearance and apparently many of the properties of Gutta-percha. Softens in warm water and becomes glutinous at the boiling point. Is soluble in sulphide of carbon, chloroform, benzole, and alcohol. Can be kneaded in water as easily as ordinary gutta. [Heckel and Schlagdenhauffen.]

GUTTA-SHEA.—Said to be the nearest approach to Gutta-percha among African products; obtained from the "Shea," "Galam," or "Bambouk" butter-tree *(Butyrospermum Parkii.)* The butter is the solid fat contained in the seeds and is used in making hard soaps. Gutta-shea is separated from the fat in the course of the soap making and is found to be present to the extent of from 5 to 75 per cent. A kind of Gutta-percha is also obtained from the trunk of the tree in small quantities. Also known as "Karite gum." [G. F. Scott Elliott, M. A., F. L. S., Botanist.]

GUTTA TERAP.—A substance closely allied to both Gutta-percha and India-rubber; used in Singapore in the manufacture of birdlime; is made from the juice of the *Artocarpus Kunstleri.* Also known as "Gutta-trap."

[Dr. D. Morris, "Cantor Lectures," 1898.]

GUTTA HORFOOT.—This is a vegetable juice sent in sealed tins from the Straits Settlements, which yields a material like India-rubber of fair quality. No way of coagulating the juice, where it is gathered, seems to be known.

[T. Bolas in Colonial and Indian Exhibition Reports, 1887.]

TALOTALO GUM.—Found in the Fiji archipelago. Comes from *Tabernoemontana Thursioni* (Baker.) The gum is hard, gutta like, and without elasticity. [Kew Annual Report, 1877.]

CATTIMANDU GUM.—This is one of the Euphorbium gums, the natives using the milk as a cement to fasten knives in their handles. Under the influence of heat it becomes soft and viscid and when dry is very brittle. It is probably about as useful as Indian gutta. Found in Vizagapatam, India. [Hon. W. Elliott.]

TIRUCALLI GUM.—This is a Euphorbium gum, from the Indian plant known as milk hedge. The milk of this plant is used for various purposes, chiefly medicinal, in India, and has been suggested as a substitute for Gutta-percha. Like Gum Euphorbium, it has a very acrid character, and the collection of it is a very dangerous operation to the eyes. When dry it becomes very brittle, but when warmed in water is quite plastic.

[India-Rubber Journal, Sept. 2, 1885.]

COORONGITE.—Sometimes known as Australian Caoutchouc. An India-rubber-like material, discovered many years ago near Salt creek, a short distance from the coast of South Australia. It was first observed in little hollows of sand and resembled patches of dried leather, but it generally occurred in the swamps. It is supposed to be of the petroleum series. Other scientific authorities in England and America ascribe to it a vegetable origin and regard the gum as exuding from a plant or lichen.

[India-Rubber Journal, Sept. 2, 1885.]

PALA GUM.—Found in Assam and Ceylon. The wood and the bark are valued in India for their medicinal qualities. The tree yields an abundant milky juice, which after coagulation acts something like Gutta-percha. It readily softens in hot water and takes impressions, which are retained when cold. Also known as "Indian Gutta-percha." Comes from the *Dichopsis elliptica*. It has been used as an adulterant of Singapore gutta for some years. It was used also as birdlime or cement and keeps well under water. Is hard and brittle when cold. The resin or crystalban is easily removed by boiling alcohol and the residue appears to be a very fair gutta. [Kew Bulletin, 1892.]

GOA GUM.—Discovered by Senbor Da Costa. It is a gum that comes from the mival-cantem, which grows wild in the Coucan district, and is also planted for hedges. Chocolate in color, softens under heat, is easily molded, and thoroughly waterproof.

MACWARRIEBALLI GUM.—A rubber gathered in British Guiana from the *Forsteronia gracilis*. From the report of the director ot the Kew gardens, to whom a sample was submitted, it would seem that, while the gum is at present unfit for use in place of ordinary caoutchouc, because of its stickiness, it might be of value in cements, frictions, and the like.

[G. S. Jenman, Botanic Gardens, Georgetown, 1888.]

CAPE CATTAMANDU.—Derived from an Euphorbia found at the Cape of Good Hope. The juice is so acrid as to give intense irritation to any part of the body with which it may come in contact. The gum has been used as an anti-fouling dressing for ship's bottoms, but is little known otherwise.

MANGEGATU GUM.—This comes from Vizagapatam and is a gum of the bastard gutta type, similar to gutta trap, and is said to come from the *Ficus Indica.*

MUDAR GUM.—This comes from an *Asclepias,* commonly known as gigantic swallow wort *(Calotropis giganteus.)* The shrub is found throughout the southern provinces of India and grows to a height of from six to ten feet. Produces a gutta-like substance, which becomes plastic in hot water, and in other ways acts somewhat like Gutta-percha. It insulates badly, but is recommended for waterproofing.

[Dr. Eugene Obach, "Cantor Lectures," 1898.]

BARTA-BALLI.—One of the best known native trees in the Guianas. The milk of this tree has usually been mixed with Balata milk and is said to give it its reddish tint. The gum when dried by evaporation is rather sticky and soft, but when precipitated in alcohol is dry and firm. Reports from England are rather condemnatory as the gum is said to absorb a great deal of water in washing, which it retains very obstinately. The same rubber, dried by precipitation by spirits of wine, is said to be very brittle. Known also as Cumaka-balli. [G. S. Jenman.]

SARUA RUBBER.—Found in the Fiji archipelago, from *Alstonia plumosa* (Labill.) Formerly collected largely, now but little comes to market. Natives take no interest in its collection. Is soft at first, but hardens after a time and becomes inelastic. Is about the color and consistency of putty. Natives collect juice in three months and it coagulates almost at once. Comes from stems and leaves. No juice in trunk of tree.

[Kew Annual Report, 1877.]

JINTAWAN.—A bastard Gutta-percha—perhaps Pontianak—mentioned by Thomas Hancock in four patents and also by Taylor and Duncan.

ZAPOTINE.—A name for a solution made from Gum Chicle dissolved in alcohol which is treated in the following manner: According to one process, Zapotine is exposed to carbolic acid gas, or to compounds containing carbon, for vulcanization. In another, in which it was claimed that it was converted into a vulcanite, the Chicle solution was combined with white lead and sulphur, and vulcanized.

MULE GUM.—Another name for Ceara rubber.

SUSU-POKO (meaning English tree milk).—A gum from a tree growing in the Malay peninsula, used in the place of Gutta-percha, after being cleansed and treated with chloride of sulphur. Mentioned by Leonard Wray in 1858.

TALAING RUBBER.—An almost black rubber which, when cut into, is white and porous presenting a honeycombed appearance, the cavities being filled with a watery fluid. It is quite tough and elastic, and appears to be of good quality. It comes from a creeper which is abundant in the Philippines, in Malacca, and Indo-China. The juice is very abundant, and is coagulated by being boiled in water.

[M. H. Pierre, formerly director of Saigon Botanical Gardens.]

CANOE GUMS.—From the bark of the breadfruit tree, which is found so plentifully in the islands of the Indian archipelago, comes a thick mucilageous fluid which hardens by exposure to the air. When boiled with cocoanut oil it makes a tough rubber-like substance wholly waterproof, and very lasting. It is used ordinarily for waterproofing seams of canoes, pails, etc. It is also used, when fresh, as a birdlime.

PICKEUM GUM.—A shrub that is said to be very plentiful in Central America and Mexico, produces a gum fully equal to African flake. The gum is gathered by cutting the shrubs and expressing the juice. A machine for this purpose is all that is needed to add another valuable rubber to the products of the countries named.

NEEN RUBBER.—A rubber-like gum said to be produced by an insect, reported from Yucatan. The insect belongs to the Coccus family, feeds on the mango tree, and swarms in those regions. It is of considerable size, yellowish brown in color, and emits a peculiar oily odor. The body of the insect contains a large proportion of grease, which is highly prized by the natives for its medicinal properties in skin diseases. When exposed to great heat, the lighter oils of the grease volatilize, leaving a tough wax which resembles shellac. When burnt this wax produces a thick semi-fluid mass, like a solution of India-rubber.

SIEBA GUM.—See Tuno.

JELATONG.—See Pontianak.

FLUVIA.—See Pontianak.

CHAPTER III.

I.—DIVISIONS IN RUBBER MANUFACTURE AND PRIMARY PROCESSES IN MANIPULATING THE GUM.

THE foremost European manufacturers of rubber goods, as a rule, make everything in the line of compounded rubber, hard or soft, and in addition often are large producers of Gutta-percha goods. In the United States, on the other hand, the tendency has been to specialize the industry and as a result it has divided itself naturally into the following general lines: Mechanical rubber goods; Pneumatic and solid tires; Molded work; Druggists', surgical, and stationers' sundries; Dental and stamp rubbers; Surface clothing; Carriage cloth; Mackintoshes and proofing; Boots and shoes; Insulated wire; Hard rubber; Cements; Notions; Plasters; and Reclaimed rubber.

The following brief description of the manipulation of rubber in these various lines is given simply because there are superintendents and managers who are experts in one line, say for example, of Druggists' sundries, but who may be wholly unfamiliar with even the machinery used in other lines.

MECHANICAL RUBBER GOODS.—This line of rubber manufacture, which is also known in Europe as technical rubber goods, embraces all the heavier combinations of India-rubber, metal, and fabric which are used in engineering and industrial lines. It covers, for example, belting, packings, hose, and special articles of almost endless variety and description.

This portion of the rubber business has always been the pioneer in the production of new compounds, new processes, and better and heavier machinery. Its manufacturers always have welcomed new grades of rubber, have been the first to utilize those that were a drug on the market, because of lack of knowledge as to their manipulation, were familiar with the uses of reclaimed rubber while yet other lines were simply considering its use, and with hundreds of compounds and cures, with a broad knowledge of industrial achievement in all lines, they have often pointed the way for manufacturers in other lines to follow, to the betterment of their goods or their pockets.

The mechanical rubber goods factory has, to begin with, the same outfit in the way of machines for manipulating the crude gum as have the other lines. Their mixing mills, however, are often heavier, and their calenders run at higher speeds, while they have in addition enormously heavy hydraulic belt presses, huge vulcanizers, and scores of special machines designed for individual problems required for their line of work alone, or perhaps for a single factory alone. The kind of vulcanization used in this work is (1) open steam heat, where the goods are buried in French talc or wrapped in fabric; or (2) dry heat, where they are confined by molds, and held in a steam press during the cure; or (3) where the goods, as in the case of belts, are molded between the platens of the press itself, while curing. Even in this line of work there are some concerns that only do special parts of it. For example, there are certain large factories that make only certain types of packings, which have a worldwide sale, and on which they are run continuously. Many of these mills also are large producers of tires.

Boots and Shoes.—The manufacture of rubber boots and shoes, although apparently a simple business, not only requires large capital but is one that has often been overtaken by disaster. It is a matter of common knowledge that, given the same compounds, the same machinery, and the same skilled workmen, no two mills are able to turn out exactly the same grades of goods. Quality is one ingredient that may or may not be added to the goods, no matter how honest the endeavor. That there are reasons for this, no one can doubt, and that the day will come when this branch of manufacture will be an exact science is probably true. That, however, will entail a definite knowledge of rubber from the moment it first sees the light as a creamy sap exuding from the tree, through every event in its life— in coagulation, transit, storage, factory manipulation, compounding, calendering, curing, its death in the service of man, and its later resurrection in the process of reclaiming.

Nor is this all. There will be a need for exact information regarding the ingredients added in the course of compounding, their relation one to another, mechanically and chemically, so long as they be joined together. This, coupled with atmospheric and climatic conditions, not to say a profound knowledge of the errors and accidents due to the ignorance, prejudice, or carelessness of the ordinary workman, constitute so complex a problem that successful manufacturers to-day feel fairly safe in frankly stating to would-be competitors that they have no need to hide their formulas, as they are but a small part of the problem.

In the complete rubber shoe plant there are found, for initial equipment, washing rolls, mixers, refining mills, and calenders such as most of the other lines employ. In addition, there are special calenders, with engraved rolls for shoe-upper work; others, also, with engraved rolls for soleing; presses for molding boot heels, sole-cutting machines, and, of course, vulcanizers. As this class of goods is cured by what is known as the "dry heat"—that is, by being confined in dry hot air for several hours—it will readily be seen that it is radically different business from mechanical rubber goods, for instance. These dry heaters are simply large air-tight rooms, fitted with steam pipes for heating, lined with tin, double walled to prevent radiation, into which hundreds of pairs of boots or shoes are run on skeleton cars, to undergo the process of vulcanization. The manufacture of rubber footwear in brief, therefore, consists in washing, drying, compounding and calendering the rubber, the cutting of the calendered sheets into various shapes for cementing over lasts in the shapes desired, the varnishing, and the dry heat cure.

To-day the ingredients used in this compounding are almost identical in all of the American mills. In Europe, however, there is a wider difference, and it would not be surprising if rubber shoe compounding experienced the same revolution that other lines have known, now that the price of crude rubber has gone so high. The last great changes in shoe compounding, which came between 1878 and 1882, were radical and of value to both manufacturer and consumer. That the present compounds are perfect, or that the ingredients used are the best, no one can affirm. Besides, as all other lines have progressed, is it not now the turn of the boot and shoe trade?

DRUGGISTS', SURGICAL, AND STATIONERS' SUNDRIES.—This part of the rubber business entails more skilful manipulation and more finesse in manufacture than almost any other line. An atomizer bulb, for example, must be graceful in shape, with delicately smooth surface, of good color, and either of the non-blooming variety or so near it that the sulphurous efflorescence will be so slight as to pass unnoticed, while in mechanical goods a length of garden hose may be of any color, may bloom until crusted with sulphur crystals, but if it "stands up to work," it is the best, and is beautiful in the eyes of the trade.

The question of colored rubber is one that has interested this branch of the business from its inception. In none other is so much white rubber made and, incidentally, none others get such good effects. This insistence by customers for white goods and by physicians for black containing no trace of lead has entailed a deal of trouble upon this trade, for the manufacturers until recently could not go into the open market and buy a high grade of white recovered rubber, while of black there is ever an ample supply, and in black goods to suit the physician he is forced to substitute a dry bulky vegetable black for oxide of lead or white lead, and then not get so good a result.

The machinery used is very similar to the equipment of a mechanical goods factory, but the scale is smaller. Washers, grinders, calenders, tubing machines, steam vulcanizers, and small steam presses are the machines used. Naturally special machines are employed in certain parts of the work, but their use is limited to a few factories and to comparatively insignificant specialties.

The feature in this trade which stands out most distinctly from other rubber lines is perhaps the manufacture of hollow work, as atomizers, syringes, breast-pumps, and a host of other balls and bulbs. The parts for these are cut from sheets of compounded rubber, cemented together at the edges, inflated to the general shape of the mold and cured in an open steam heat. In order that the ball may perfectly fill the mold during the cure, a few drops of water or a little ammonia are put inside of it which, swelling under the heat, develops pressure enough to perfectly shape it and add to its outer surface the finish found on the inner surface of the mold.

The difficulties that manufacturers in this line experience in making perfect goods are legion, as they are in other lines. They are added to by the fact that the trade, as already indicated, demand articles of beauty from a gum that was designed for utility solely. A trace of black in a white compound may spoil hundreds of dollars worth of goods, nor can such trace be rubbed off, scoured out, or eradicated, after vulcanization. Hence, the whites, blacks, reds, and other colors must be mixed in separate mills, and the trimmings and scraps kept sedulously apart.

Pure gum—that is, rubber compounded only by sulphur or some other vulcanizing agent—is also largely produced in this line. For example many make what is known as dental dam, the pure sheet used by dentists. This is generally a sulphur compound cured in open steam. Certain manufacturers, however, practice the vapor cure with good success in making these goods. This cure gives a beautiful finish, but if it is not done with great skill it is disastrous to both the workman and the goods.

Dental dam, surgical bandages, and stationers' bands represent the highest priced and least compounded goods, while stopples, erasive rubber, and common tubing represent the other extreme. Between the two is a latitude that allows of a variety of combinations and compounds that no man can number.

CLOTHING, CARRIAGE CLOTH, MACKINTOSHES, and PROOFING.—This business may be handled, in a measure, as the mechanical goods business is; that is, the gums mixed by heat on ordinary mixers, and then spread by calenders on the fabrics which give the articles their strength. This is the manner in which rubber surface clothing is run. The machinery is simple, since, in ciothing, the parts are cemented together and cured in dry heat. In carriage cloths, after calendering, the goods are grained on embossing rolls, varnished, and run into a dry heat.

The Mackintosh and proofing business, however, is somewhat a departure from this. Here the gum, after mixing dry, is usually put in churns with a cheap solvent, and reduced to a solution. It is then applied to the cloth with a knife spreader.

For double-texture work, a simple doubling machine brings two surfaces together. A portion of the business that has divided itself from the rest, is what is known as proofing for the trade. Here manufacturers simply coat the cloth and sell it to others, who make it up into garments, or anything in fabric or rubber for which there may be a call. The mackintosh manufacturer to-day not only is familiar with a great variety of rubber gums and ingredients used in compounding, but is also an expert in fabrics, as his business is really closely akin to the tailoring business.

Pneumatic Tires.—Although the tire business seemed at first to be a natural part of the mechanical rubber goods business, it really proved itself, later, to be a business wholly distinct from it. Even the large manufacturers of mechanical goods who began tire making on a considerable scale, keep this part of their business distinct from other branches as a rule, running it as an entirely separate department. Aside from this, large concerns have sprung up that manufacture nothing but tires, and although some of these use their scrap and refuse in the manufacture of certain mechanical goods, they do not all find it profitable.

The general machinery used in making tires is the same that is used in the work of preparing rubber in the other lines. There are two general classes of tires manufactured, however: those that are molded, and those that are made in such a way that they can be wrapped for the process of vulcanization. Wrapped goods, of course, are cured preferably in an open heat. In the one case the tires are cured in presses, sometimes in nests of molds, and sometimes in vulcanizers. Molded tires are cured under pressure, exactly as the atomizer bulb is in the druggists' sundries line. Various ingenious and valuable processes in special machines have been invented, and are now in use in this industry. A minor industry that has grown up in connection with the tire business, and that has increased the practical knowledge of the uses of rubber wonderfully, is that of tire repairing.

A part of the tire business that is of great interest is the making of the solid or cushion molded tire used on vehicles. A very large business is done in this, the work being a simple process of mixing the prepared compound, forcing it into shape through a tubing machine, and molding in an open steam heat. A tire now coming into use that is going to develop a very large business is the big pneumatic tire used on various types of automobiles. The knowledge gained through the manufacture of pneumatic bicycle tires, which, by the way, was one of the hardest problems that the rubber trade ever solved, has proved wonderfully effective in developing the skill necessary to make this heavier and more important article. This tire, like the bicycle tire, is built up of frictioned duck, with an outer coating of high-grade rubber carefully vulcanized. While a variety of compounds undoubtedly will be used in its manufacture, it is hardly possible that any manufacturer will be able to sell a very low grade of goods. In other words, the life of the tire is so important, and the purchaser so anxious for a good article, that adulteration or cheapening to any great extent is not a present danger.

INSULATED WIRE.—The manufacture of insulated wire, either with India-rubber or Gutta-percha insulation, is a line that is more distinctly apart from other portions of the rubber business than almost any other. For Gutta-percha, the general machinery used is described in the chapter on that gum. Where India-rubber is used, the crude gum is treated in the same way as in mechanical goods. It may be forced over the wires by tubing machines, or welded together in strips that are run between grooved rolls.

Braiding machines are also a part of the outfit for weaving the protective covering, and the wire is usually wound on huge drums and vulcanized in open steam heat. Polishing machines, testing machines, and various mechanical contrivances are, also, a part of this equipment. The line of compounds used is one adapted almost wholly to this industry, and embraces a great variety of ingredients and gums that are treated specifically under their special heads, elsewhere in this book.

MOLD WORK.—A part of the rubber business that belongs either to the mechanical or the druggists' sundries line has, during the past few years, detached itself from the rest, so that to-day many large factories are run simply in producing small mold work. They have the usual equipment of rubber machinery, special appliances for filling and emptying molds, and the usual aggregation of hard and soft metal molds that run into thousands of dollars in a short time. The extent to which this business is carried may be imagined when it is known that one company runs 80 presses on this work, and many have from 20 to 50 in constant service. When it is remembered that very rarely are two compounds exactly alike, it will be seen that, in this line also, the expert compounder has a wide field for thought and experiment.

Hard Rubber.—In spite of the hundreds of substitutes for vulcanite, or hard rubber, that have been produced, the demand has in no way fallen off, and these mills are running full to-day on the production of this semi-metal. The old fashioned compound, consisting of 2 pounds of India-rubber to 1 pound of sulphur, is still in use in certain goods. Modern progress and chemical knowledge have, however, added a great many compounds for specific uses, so that almost any degree of quality or hardness or price is now furnished on call.

The business, primarily, is a simple one, the hard rubber machinery being like that used in other lines. In the manipulation of the gum for vulcanization, and in its finish, however, special machines are necessary. The finishing machines are lathes, saws, buffers, etc., somewhat similar to what might be used for turning hard wood. The mechanical factories often do a little in hard rubber in the line of valves, and the druggists' sundries mills often make their own syringe fittings, but the bulk of the business in America is done by mills that make only vulcanite the year around.

Cements.—Many rubber factories are run wholly on this line of work, the gums being mixed as in a general rubber business, put into solution in churns, and sold by the barrel for an infinite variety of purposes. Hundreds of different formulas are in use for cements sold for general and specific purposes. The leather shoe business, for instance, calls for a dozen or more special cements. The bicycle business has need for a great many grades of what are known as tire cements and what are known as puncture fluids. The latter, however, do not really belong to the cement business. Stickiness, waterproof qualities, durability, and cheapness in their goods are sought by all cement manufacturers, and, in order to secure these qualities, skill is demanded in compounding in no way inferior to that shown in other lines of rubber work.

Dental and Stamp Rubber.—The manufacture of unvulcanized gums for the use of dentists and rubber stamp manufacturers is an industry apart from other lines, and one that has assumed quite large proportions. The rubber is compounded and sold by the manufacturer, and cured and finished by the dentist or rubber stamp manufacturer. In stamp work the rubber is compounded for soft rubber and many hundreds of tons are sold during the year, while of course the dental rubber is so mixed that under the cure it becomes vulcanite of the color desired. The machinery for this work consists chiefly of washers, mixers, and calenders.

NOTIONS.—A department of the rubber business little known is that which takes in such work as waterproof dress bindings, dress shields, childrens' aprons, diapers, etc. Several large factories manufacture these goods, mixing their rubber by the usual dry process, coating it on calenders, and having special machines for forming and curing the goods in their special shapes. In the manufacture of dress shields, the vapor cure is often practiced very successfully. The rubber manufacturers of this class are not by any means inexpert compounders. They have also, perhaps, gone as far as any in deodorizing rubber goods, so that the smell of the gum or any compounding ingredients is wholly done away with.

PLASTERS.—There are few factories that keep wholly to this line of work. It is perhaps as simple as any part of the rubber business, a fair grade of rubber being washed, dried, and mixed by the usual methods, and calendered upon the fabric that forms the base of the plaster. These goods are not vulcanized, of course. Though a variety of gums and medicaments is used in this compounding, the range is probably smaller than any other line of rubber manufacture.

RECLAIMED RUBBER.—In the United States nearly a dozen mills are employed in the reclaiming of waste rubber, such as old boots and shoes, hose, tires, etc. In this business are used crackers, sheeting mills like ordinary grinders, and, indeed, general machinery not dissimilar to that used in a mill where crude rubber is compounded. They have in addition, however, lead lined tanks for acid treatment, vulcanizers or, better, devulcanizers, huge vats for washing, magnets for removing metal, sieves, and the like. This branch of the rubber business is not supposed to be deeply interested in compounding, in spite of the fact that it is sometimes suggested that earthy matters and heavy adulterants do find a use in reclaiming mills.

II.—THE WASHING, MIXING, AND CALENDERING OF RUBBER.

THE primary process that rubber undergoes when it enters a rubber mill after weighing, is washing. As a rule this is done with clear water. At the same time, certain acids, alkalies, and foreign substances that are contained in the rubber are not easily soluble in water, and yet may be easily removed. The first thing to do, therefore, is to know what is to be expected in various grades of rubber. Perhaps there is no better way to get a bird's eye view of what the washer might wish to remove from the gum than by briefly cataloguing the different substances used by the natives in coagulating the juice.

There is no question but that the differences between varying grades of rubber, besides being due to a somewhat different chemical composition, are also due in a measure to varying methods of collection and coagulation of the sap. It is undoubtedly true that no one method of collection would be best for all kinds of rubber gathered, even if it were possible. At the same time, it is of interest to the practical rubber manufacturer to know pretty nearly what systems are pursued, and particularly what ingredients are added to the sap, to produce coagulation, as the presence of certain residues may affect his compounds.

SMOKING rubber is the system with which the world at large is most familar, and is practised in the Amazonian forests in the collection of Para gum. Several kinds of palm nuts are used to produce a thick smudge, but those ordinarily used are from the Urucuri palm *(Attalea excelsa.)* This smoke has been found by analysis to consist mainly of acetic acid and creosote, the latter being a well known preservative of rubber. Fine Para rubber is nearly always smoked in this way. Coarse Para is air dried. Ceara rubber is also, to a certain extent, smoked in the gathering, the palm nut used being that of the *Eucturbe edulus.* There is also a kind of gum tree found in the forests of the Isthmus, and where it is impossible to get palm nuts, its wood is used for the coagulating smoke.

ACHETE JUICE.—A native process for coagulating the sap of the rubber tree, which prevails throughout Central America, involves the use of an alkaline decoction made from the juice of a plant called "achete" or "coasso" (*Ipomoea bona-nox*, Linn., and also *Calonyction speciosum*). This is combined with rubber milk in the proportion of 1 pint to 1½ gallons of the latter. During coagulation the vessels are often heated from 165° to 175° F. After coagulation, the rubber is dried for twelve or fourteen days. The kinds of rubber coagulated in this fashion are Mexican, Nicaraguan, and in fact almost all of the rubbers that come under the head of Centrals and are obtained from the *Castilloa elastica.*

SULPHUR FUMES.—According to James Collins, rubber of the Para varieties is sometimes exposed to the action of the fumes of melted sulphur, which affects coagulation. This process, however, is very rarely followed.

COYUNTLA JUICE.—This is an astringent juice made from the Mexican weed of that name. When the rubber milk is gathered. it is placed in earthenware vessels and whipped with the weed, which causes coagulation. The Mexican rubber known as Tuxpam is treated in this way.

MACHACON JUICE.—Cartagena rubber, which is gathered carelessly, is coagulated in a hole in the ground by the addition of the juice of the root of the "machacon"—a strongly alkaline solution.

NIPA SALT.—A salt obtained by the burning of the plant known as the *Nipa fructicans.* Is used in the coagulation of Borneo rubber.

LIME JUICE.—Lagos rubber and some other African sorts are coagulated by the addition of a little lime juice, which is added as the sap flows from the vine.

ALUM.—This is used all through the Isthmus of Panama, in coagulating Accra rubbers, and other African sorts. Pernambuco rubber is also treated with a water solution of alum, as is the Nicaraguan at times.

SALT.—Many kinds of low-grade rubber are coagulated by the addition of salt or brine. Borneo, for instance, is coagulated in that way. Madagascar rubber receives a treatment of salt-water. Mangabeira rubber is treated with a mixture consisting of 1 part of salt to 2 parts of alum. Nicaragua rubber is also often coagulated with salt.

LIME.—A final process in the coagulation of rubber in India

is the washing over with lime. Collins also mentions the use of lime in connection with the coagulation of Para ruber.

Soap and Wood Ashes.—The medium grade rubbers all through Central America are often coagulated by the use of soap, and where that is not plenty, of a strong lye from wood-ashes.

Spirits of Wine.—This is used sometimes in the coagulation of Balata.

Torres System.—In addition to the natural methods described above, there are several that give some evidence of an intelligent study of the sap and the substances best adapted for this work. Under the Torres system a liquid is made by a secret formula, from the roots and fruits of certain South American palms, which, when added to the sap, preserves it from curdling, so that it will keep for weeks. It can thus be transported to a convenient place for smoking.

Helfer Process.—This consists of the addition of a solution of acetic acid, and is based on the knowledge derived from the analysis of the smoke of the Urucuri nuts.

Centrifugal System.—Another form of coagulation, that has recently been tried with considerable success, is the using of a centrifugal machine which removes the watery contents from the gum, and produces a marvelously clear elastic rubber.

Heat, Air, Sunlight.—Various rubbers are coagulated simply by the exposure to slight artificial heat, to the sunlight, or merely to the air. Such are the coarse Para rubbers, certain of the Centrals, African, and East Indian rubbers. Fiji rubber is coagulated in the mouths of the natives, and Angola rubber on the arms and breasts of the natives.

The very first manufacturing process in the manipulation of rubber of any kind, and for any use, is that of the cleansing. This is usually done by passing the gum again and again between corrugated rolls, while fine streams of water remove the various impurities that are exposed by the tearing action of the rolls. These impurities are bits of vegetable substances, earth, sand, acids, and alkalies. The old type of washer for removing these was a couple of corrugated rolls 6 or 8 inches in diameter, and 12 or 14 inches in length. Modern methods, however, have introduced larger rolls, until to-day one machine, when it is the highest type of three-roll washer, will cleanse enough gum to keep a huge factory busy.

Some rubbers are so full of sand that it is almost impossible to remove it wholly. For this purpose is used a tub with a false bottom made of fine wire, and also with a stirrer. The thimbles, for instance, after being run through the washer, are put in the tub without any attempt at sheeting, and stirred until a large portion of the sand is removed.

Another type of washer is one that is quite similar to a paper engine; in fact, paper engines are often used in rubber washing. The special value of this type is that the rubber in its movement about the tub is floated more or less, and the sand and earthy matters sink to the bottom, while the bark and vegetable matters can be seen and easily removed.

Certain manufacturers, following Austin G. Day's ideas, have used alkaline solutions in washing certain gums, to neutralize the vegetable acids, and it is a question if it might not be as well to use dilute acids to neutralize the strongly alkaline qualities of gums that go through certain kinds of coagulation. Some factories also examine the coarser grades of gums chemically, and give them a treatment to remove odor. As a rule, however, manufacturers rush them through the washing machines, sheet and dry them, and get them into the mixing mills as soon as possible.

The drying of rubber, according to earlier practice, required a great deal of time. It was the boast of more than one rubber mill that no Para rubber was used by them until it had been dried for a year. The manufacturers of mechanical rubber goods were the first to break away from this tradition. In many cases they found, when there were rush orders on hand, that they must put on their mills gum that was practically just off the washer, and mix it, or else lose orders. Of course, they were forced to get most of the moisture out, or neutralize what was left, and they learned incidentally that they got a stronger compound with the green gum than with the "seasoned," whence the belief grew up that the months and years of drying was not necessary, as had before been supposed. In addition to this, some of them learned that long drying meant oxidation on the outside, or the turning of rubber into resin, which further increased their doubt of the wisdom of the slow drying process.

These thoughts once entertained, it was not long before various plans were introduced into the drying, for hastening the removal of the moisture. The simplest of these, of course, was artificial heat, and the presence of a fan for removing the moisture laden atmosphere. Later developments have brought about a process for drying rubber very cheaply at quite a high heat, lasting only a few hours, that gives it to the man who runs the mixer, hot from the dryer, and that wholly does away with the expensive process of breaking down. This latter idea, is to some, of course, as revolutionary as was the first thought of quick drying, but that it is wholly in the line of progress, is proved by the fact that it has been used for a number of years in one large factory whose goods stand very high.

The milling of crude rubber is simply putting the dry rubber which is found in a tough, intractable sheet, on hot rolls, and running it until it gets to be a softened homogeneous mass. The gum, when this is accomplished, is ready for mixing. These mixing rolls are run at different speeds and are called friction rolls, and the various adulterants and ingredients that are to be incorporated with the rubber are pressed into a softened gum by their revolution.

No general rule can be laid down for mixing in all lines. An expert compounder knows that certain gums should be mixed on cool rolls, and others under considerable heat. His knowledge of specific compounds teaches him to hasten mixing in many cases where another, without skill, would require very much more time to get the same result. In some cases one ingredient is put in with the others, in some, it is necessary to put it in last. Some have dissolved substances that would make the rubber stick to the rolls like glue unless they are put in at just the right time; others have so large a proportion of earthy matters that, unless the gum is humored, it apparently will not take them in, and so on. Each line of work and, in fact, each factory has its own special methods, and often one or more skilled mixers who can handle compounds that none of the others seem to be able to do anything with.

The use of the calender is simply to sheet the goods so that they may be easily made into the desired forms. The simplest form of calender is a mixing mill with the key that normally holds one roll in place withdrawn, so that both run by even motion, which is used in many small factories where nothing but molded work is made.

The modern sheeting calender is ordinarily a three-roll machine. It is sometimes made with four rolls, however, and these rolls may be almost any size, the widest for rubber work being little less than 80 inches. No little skill is required for running the calender on a variety of stocks, nor can any general rules be laid down for calender work. This is proved by the value that is set upon good calender men, and by the difference that there is between the work of a good one and a poor one. There are as many different kinds of calenders as there are patterns of mixing mills. A sheet calender has smooth rolls, and is for running absolutely smooth goods. In shoe work there are engraved rolls, pebbled rolls, and soleing calenders engraved in the likeness of the shoe sole. The carriage drill business has embossing calenders, and so on. A type of calender that is useful in most lines of work is known as the friction calender, the rolls in which, run at uneven speeds, drive the gum deply into the fabric.

Where India-rubber is handled in solution there is used in place of the calender a spreading machine, known under various names of "Yankee flyer," "English spreader," "Doughing machine," etc. In this a sheet of rubber is spread on the cloth by being placed on an endless apron of the fabric, the apron running over the roll against which hangs a heavy knife. A very thin coating of the rubber solution is constantly scraped off this surface, which then passes over hot drums or steam chests, evaporating the solvent.

CHAPTER IV.

VULCANIZING INGREDIENTS AND PROCESSES.

WHILE Charles Goodyear's patents for the vulcanization of India-rubber by the use of sulphur and heat were in force, a marvelous amount of ingenuity was shown in the attempts to accomplish the same results by the substitution of other ingredients for sulphur, either with or without the use of heat. These experiments and inventions embrace vulcanization, by means of chlorides, nitrates, nitrites, fluorides, bromides, iodides, and phosphorets of about all of the common earths and metals, and also many gases such as sulphurous acid gas. The majority of these experiments have been lost sight of, partly because the Goodyear process is now open to the world, and partly because, for the majority of goods, the sulphur and heat cure is not only the cheapest, but the easiest to accomplish. It may be well, however, to review and record the experiments in this line, as there is no doubt that for special lines in rubber manufacture many of them have a great suggestive value to-day.

One of the very first ingredients to which inventors and experimenters turned their attention was zinc. The veteran rubber manufacturer, the late Jonathan Trotter, described a process for preparing a vulcanizing material which he called hyposulphite of zinc. It was made from a solution of caustic potash saturated with flowers of sulphur and then treated with sulphurous acid gas. This solution he mixed with a saturated solution of nitrate of zinc, forming the precipitate that he desired. He used 3 pounds of hyposulphite, to 10 pounds of rubber, curing from 3 to 5 hours, at 260° to 280° F.

Another American, E. E. Marcy, some years later patented a compound of hyposulphite of zinc and rubber which is apparently almost identical with Trotter's discovery, although he disclaimed similarity, and also made public the process in which he used a combination of hyposulphite of zinc and sulphide of zinc, the compound being 2 pounds of rubber, 1 pound sulphide of zinc, 1 pound hyposulphite of zinc, and other ingredients as deemed necessary. These goods were of a beautiful white color, were said not to bloom, and did not need the sunning process then in use. At the same time they depended upon sulphur and heat for whatever vulcanizing was accomplished.

Another attempt to get a good substitute for sulphur was in the production of what is known as sulphite or hyposulphite of lead. James Thomas describes at length a compound in which he mixes hyposulphite of lead and artificial sulphide of lead in equal proportions, his compound being for vulcanization, 2 parts by weight of India-rubber and 1 part of the vulcanizing material.

Following this thought, came E. E. Marcy again, who mixed sulphide of lead and carbonate of lead in the proportions of 2 parts of sulphide of lead, 1 part carbonate of lead, and 2 parts protoxide of lead in place of the carbonate.

Then Oscar Falke and Albert C. Richards brought out a compound consisting of 6 parts India-rubber, 2 parts sulphide of antimony, and ½ part sulphite of soda, curing at 270° to 280° F.

A. K. Eaton, in no uncertain terms, disclaimed vulcanization by the use of free sulphur, but claimed to be the first to use sulphide of manganese. He also gave a formula for making it, which was by mixing intimately 44 parts of peroxide of manganese with 32 parts of sulphur, and exposing the mixture to heat in a covered crucible. He vulcanized several hours, from 250° to 310° F.

George Dieffenbach claimed sulphite of alumina as an ingredient which, in connection with heat, would bring about vulcanization. He used this in a compound for a dental rubber, which had for its basis India-rubber, amber, linseed oil, sulphide of cadmium, oxide of tin, vermilion, and pulverized feldspar.

Charles T. Harris cured India-rubber by combining it with an artificial sulphide of bismuth, which he explained as being the artificial tersulphide, or polysulphide of bismuth. He describes this as being a heavy black powder, and the compound which he advised for soft rubber was 100 parts India-rubber, 75 parts carbonate of lead, and 12½ parts polysulphide of bismuth, cured in a dry heat at 245° F. for 1½ hours.

The veteran Henry W. Joselyn discovered that shale—an earth that is very plentiful in New Jersey—combined by heat with sulphur, formed a sulphide which could be used in curing rubber, and hastened to patent it.

Andreas Willman, with more originality, brought out a process for combining India-rubber with "anhydrous chlorides, sulphates of alkalies" and powdered coke or coal, and claimed that his best result came from chloride of ammonium and coke. His compound was made up of litharge, lampblack, and powdered coke, in connection with from 2 to 10 per cent. of his vulcanizing mixture.

Edwin L. Simpson formed a vulcanizing compound by mixing benzoin gum with pulverized sulphur, and boiling it in linseed oil. It was used in a dry heat, the compound being 1 pound of India-rubber, 2 ounces vulcanizing compound, 8 ounces litharge, and 8 ounces whiting.

J. A. Newbrough manufactured a vulcanizing material which he called acid resin, made of turpentine and sulphuric acid. This he incorporated in India-rubber in the proportion of 6 ounces of acid resin, to 1 pound of India-rubber, and cured at 300° to 320° F.

The use of selenium as a curing agent was discovered by E. E. Marcy, while connected with Horace H. Day, then prominent as a rubber manufacturer. He advised the use of equal parts of India-rubber and powdered selenium, and, to produce a glossy finish, he added selenium carbonate and whiting.

At the same time there were many other inventors who were experimenting with processes that were somewhat in the line of the well-known Parkes cold-curing process. For example, it is a matter of history that the late Joseph Banigan, early in his career as a rubber manufacturer, cured wringer rolls by an acid process.

Dubois C. Parmelee invented a process which he called "hermizing," to distinguish it from curing or vulcanizing, instead of the Parkes process, in which the solution of chloride of sulphur and bisulphide of carbon were used. He recommended briefly a solution made as follows: 10 pounds of coal-tar naphtha, in which was dissolved 1 pound of sulphur. Into this solution he passed dry chlorine gas until it assumed a fine yellowish-green color. This solution he used as a dip for such goods as would be cured by the acid treatment. Parmelee also claimed the discovery of a solution made of coal-tar naphtha, bisulphide of carbon, and a solution of sulphur in bromine, mixed with this.

H. A. Ayling patented a cold curing process in which carbon

spirits, one of the petroleum series, was mixed with chloride of sulphur, instead of the usual bisulphide of carbon.

Referring again to the suggestions of chlorine in the working of rubber, R. F. H. Havermann reduced India-rubber to a solution and subjected it to the action of chlorine. He also, in a later patent, described the washing of the chlorine out of the rubber with alcohol, and the addition of ammonia and lime, the result being, according to his specifications, a white hard rubber.

Working in the same line, John Helm, Jr., dissolved India-rubber in benzine and mixed it with liquid chlorine in the proportion of 12 ounces of chlorine to 1 pound of gum. His claim was that he could get rubber of any color and of any degree of hardness by this process.

In the line of hard rubber manipulation and vulcanization, Mr. Meyer (connected with the India-Rubber Comb Co.) patented a process for curing vulcanite in a vessel wholly or partly filled with water, the water in which the rubber was contained being in a tight receptacle, and the heat being raised above 300° F., the pressure of the surrounding steam keeping it from vulcanizing. This obviated the danger of burning, and was of great value in the production of certain goods.

While these and other inventors were trying to cure rubber without sulphur, and without interference with the Goodyear patents, certain others were at work on other gums. For example, John Rider, who was at the head of a Gutta-percha company, produced what he called mettallothyanized Gutta-percha. In this, he first heated the Gutta-percha, then mixed 3 pounds of hyposulphite of lead and zinc with 8 pounds of gum, and sometimes added also a little Paris white, or magnesia. He then put the compound from 2 to 10 hours in a dry heat and cured it at 280° to 320° F.

John Murphy changed this compound somewhat, by advising the incorporation of sulphur in the proportion of 2 to 6 ounces of sulphur, to 10 pounds of Gutta-percha. This sulphur, by the way, obviated the preliminary heating of the Gutta-percha, which was supposed to volatilize the ingredients that had before rendered it unvulcanizable.

A curious process for the manufacture of hard rubber was also brought out by William Mullee. In this, just as soon as the rubber was washed, the sheets were immersed in the sulphur bath, heated to 220° F. The water and other impurities in the rubber were said to be extracted by the action of the heated sulphur. After boiling 30 minutes, the sheets were removed with tongs and washed to prevent crystalization. They were then subjected to the same process a second time. The rubber was then compounded in the old fashioned way, on rolls, the proportions being 17 to 24 ounces of sulphur to 16 ounces of rubber. The claim for this was, that the compound when cured was tougher than any others ever known.

William Elmer prepared what he called "elastic selenide of caoutchouc." He first dissolved the India-rubber in bisulphide of carbon, placed it under pressure, and heated gradually. When brought to about 300° F., the liquified selenium was put into the apparatus drop by drop, the solution in the meantime being kept in constant motion. This elastic selenide he claimed to be semi-fluid which, when evaporated, possessed all the characteristics of India-rubber.

The Parkes cold-curing process is so widely known as to require but a word. It is based on the invention of Alexander Parkes, and depends upon the faculty that chloride of sulphur has for vulcanizing India-rubber. (See Chloride of Sulphur.)

A curious process, similar to that of Parkes, is Caulbry's process, by which it is claimed rubber can be vulcanized at ordinary temperatures, by using an intimate mixture of chloride of sulphur and dry chloride of lime. During this mixture, and when the smell of the chloride of sulphur will be noticed, the temperature of the mixture will rise, the mass becoming plastic by the softening of the sulphur. If a mixture of this kind, in which sulphur is in great excess, is added to the solution of India-rubber in bisulphide of carbon, the rubber will be vulcanized at an ordinary temperature, or perhaps with a slight warming. Chloride of sulphur used pure is too corrosive in its effect on India-rubber; it is therefore reduced in all cases. Only thin articles can be vulcanized in this way.

A recent patent taken out in England by Edmond Garnier relates to the vulcanization of India-rubber by the use of alum. Alum processes for curing in the past have not been very successful. This patent, however, has some novel features. It calls for particularly dry alum treated with a solution of terebinth of benzol and shellac, or some similar gum. In use he takes 8 ounces of alum and a solution composed of 1 part gum and 20 parts benzol. He mixes together the ingredients that are usually employed in the manufacture of rubber, specifying 3 pounds of whiting, 1 pound barytes, 8 ounces lime, 1½ pounds oxidized oil, and 8 ounces of India-rubber. When these have been thoroughly mixed together and specially treated, alum is incorporated with them and well compounded, being passed through the mixing rollers cold. It is then calendered.

AMORPHOUS SULPHUR.—The fusing of 1 pound of sulphur with 4 ounces of Canada balsam produces what is known as amorphous sulphur, which is said to cure rubber so that it will have no tendency to bloom. The preparation has a very pungent sulphurous odor. Patented by Dr. F. Wilhoft, of New York.

ARTIFICIAL SULPHURET of LEAD.—There are several combinations of lead and sulphur which may be produced artificially. That one containing the most sulphur has a composition of 13 per cent. of sulphur and 86 per cent. of lead. Its specific gravity is about 9.4. In color it is black. It melts at a strong red heat. The other sulphur compounds of lead have much less sulphur, one containing but 9 per cent. and the other only 4 per cent. What is known as hypo-sulphite of lead is a mechanical mixture of the above first named, with a suitable percentage of sulphur to effect vulcanization. It is also known in the rubber trade as "Eureka compound" and "Burnt hypo." These compounds when pure—that is, when free from adulteration—are of great value. They produce goods that are jet black and have little odor and are free from bloom. They are reckoned as the safest vulcanizing agents, as it is almost impossible to burn goods that depend upon their presence for cure. They are used in either dry or wet heats.

BARIUM SULPHIDE is prepared from heavy spar by making a dough of it with charcoal and oil and subjecting it to a white heat. Sulphides of the alkaline metals, potassium, sodium, calcium, and barium, will vulcanize rubber, whence the term "alkalised rubber."

BROMINE.—A heavy deep red volatile liquid, possessing a most peculiar and unpleasant odor, and giving off vapors most irritating to the air passages and lungs. It's very name means stench. It has a powerful action upon most organic bodies, coloring animal matter brown, while it bleaches coloring matters, dyes, etc. Its specific gravity is 3.18. A piece of sheet rubber dipped into bromine is vulcanized instantly. It is somewhat soluble in alcohol, and very soluble in ether, bisulphide of carbon, chloroform, etc. Messrs. Newbrough and Fagan filed two patents in the United States for the use of bromine in vulcanization, both with and without iodine. By adding to iodine ½ its weight of bromine, proto-bromide of iodine is formed, which is said to combine with India-rubber and produce a hard compound on being exposed 1 hour to a temperature of 250° F. To prevent the forming of an explosive the iodine and bromine were separately treated with oil of turpentine to which had been added a quarter of its weight of sulphuric acid. It was then mixed with the gum in the proportion of 2 pounds 11 ounces to every pound of gum. Bromine was also used alone by these inventors, the material after molding being plunged into the liquid, and left there long enough to harden. To prevent the hardening of the material, while in the bath, chloroform or any other solvent of rubber was added in the proportion of 1 part to 9 parts of bromine; in other words, the rubber vulcanized in the air after its withdrawal from the liquid.

Chloride of Sulphur.—Sulphur and chlorine form three compounds, the monochloride, the dichloride, and a tetrachloride of sulphur. The substance usually used in the arts is the first named or a mixture of the first two. It is an oily liquid of the specific gravity 1.7, and boiling at 239° F. It has a pungent smell and decomposes on contact with water or watery vapor. Pure chloride of sulphur is of an orange yellow color of great density. It fumes strongly when exposed to air, throws off the vapors of hydrochlorine, and is quite poisonous, severely attacking the mucous membranes. It is widely known as the active agent in Parkes's cold-curing process, where it is used in connection with bisulphide of carbon. A common formula for this is chloride of sulphur, 1 part by weight, bisulphide of carbon, 30 to 40 parts by weight; immerse from 60 to 80 seconds. In the manufacture of balloons and toy balls, the solution is a far weaker one. That for the outside dip is 10 parts of chloride of sulphur to 100 parts bisulphide of carbon, while for the inside it is 16 parts chloride of sulphur to 1,000 parts bisulphide of carbon. When it was common to cure proofed cloth by the cold process, it was done by wetting its surface with a mixture of 5 to 10 parts of chloride of sulphur, dissolved in 100 parts of bisulphide of carbon, then running the fabric over heated drums to evaporate the mixture. In the sulphurization of oils for rubber substitutes chloride of sulphur plays a most important part, nearly all of the amber and white products being produced by its use. It also has a curious effect upon bastard gums, giving some of them temporarily the elasticity and appearance of high grade rubber.

GOLD BRIMSTONE.—See Sulphur.

GOLDEN SULPHURET OF ANTIMONY.—This is prepared from black antimony by boiling it with caustic soda and sulphur for some time. The liquid is then clarified by filtration or settling and the clear part treated with a dilute acid, preferably muriatic or sulphuric. A golden yellow precipitate is formed which should be well washed in water, and dried at not too high a temperature in a darkish place. The results of this operation well carried out are constant and the composition should be: Antimony, 60.4; sulphur, 39.6. Golden sulphuret of antimony heated in a tube will give off sulphur which will deposit on the cool sides of the tube away from the flame and the residue will turn black, being indeed the black sulphide of antimony. All samples of this compound should be tested for free sulphuric acid by shaking up a little of the powder in a test tube with cold or hot water, and testing the water afterwards with some barium chloride and blue litmus paper. A white cloud in the first place and the reddening of the paper in the second place indicate the presence of more or less free sulphuric acid. Golden sulphuret prepared with muriatic acid will not respond to the first test, but will to the second.

GOLDEN SULPHURET OR ANTIMONY RED (penta-sulphide) is used more largely than any other form of antimony in rubber work. It is frequently adulterated, sometimes with carbonate of lime, oxide of iron, or oxide of antimony, all of which tend to harden the rubber. Also called Orange Sulphide of Antimony. Properly used, this ingredient produces some of the best effects found in vulcanized rubber, in color, texture, and durability. It should never be mixed on a very hot mill, should be sheeted and placed in cooling racks if it is not to go right to the calender, and should be cured in as low a heat as possible. The ideal result will be of a golden yellow color, with a very slight bloom, if any. It is used only in high cost goods.

Honeycomb Sulphur.—A vulcanizing compound made by boiling a pound of sulphur, and two ounces of benzoin gum together, 1 pound of this material being mixed with a quart of boiled linseed oil.

Hypo-Sulphite of Lead.—See Artificial Sulphuret of Lead.

Iodine is manufactured from seaweed and is a black-gray substance occurring in small shining scales. Its specific gravity is 4.94 and it fuses at 239° F., giving off violet vapors. It is readily soluble in alcohol, benzol, chloroform, and sulphide of carbon. In addition to the formula given under the head of bromine, Messrs, Newbrough and Fagan patented the combination of iodine and sulphur. In this the sulphur was boiled in turpentine, and the oil decomposed and deposited with the sulphur at the bottom of the vessel was used in the operation, after being washed in dilute sulphuric acid, and dried. The iodine was treated in the same manner to prevent explosions. Equal proportions of the two were melted together and incorporated in the proportions of 2 ounces 5 drams, to 1 pound of rubber. After shaping, the articles were put in a vulcanizer and during the first fifteen minutes exposed to a dry heat, gradually increasing to 320° F., remaining there 5 minutes, then dropping rapidly to 250° F., and continuing for an hour.

Liquid Chlorine.—Chlorine is a greenish yellow gas at all ordinary temperatures. It has strong bleaching properties and also a very bad smell and action upon the respiratory passages. Under a pressure of 127 pounds to the square inch at 60° F., chlorine condenses to a yellow liquid, having the specific gravity of 1.33. This liquid, however, is unknown in the arts. It is probable that either a solution of the gas in water or as sulphuric chloride in bisulphide of carbon is meant. It has been contended that chloride, especially in the last-named solution, is the really active agent in curing caoutchouc. Chlorine cannot, as a rule, destroy mineral colors or blacks produced by carbon. Helm claimed that he was able to produce white hard rubber by incorporating chlorine with the mass.

LIVER OF SULPHUR.—This is really penta-sulphide of potassium, and is obtained by mixing carbonate of potassium together with sulphur. It is called Liver of Sulphur on account of its brown color. As it is quite volatile it should be kept in well closed glass vessels. The fluid for vulcanizing purposés is a concentrated solution of the penta-sulphide, about 25° Baume being right for use. To cure with it the liquid is brought to the boiling point in a porcelain vessel, the articles to be vulcanized being immersed in it. This is known as Gerard's process and is said to be inexpensive and perfectly safe.

MILK OF SULPHUR.—Another name for what is ordinarily termed precipitated sulphur. It is fine, light, and grayish white in color, but is often adulterated with sulphate of lime. It should be kept in a dry place, as it has an affinity for moisture.

PENTA-SULPHIDE OF ANTIMONY.—The chemical name for Golden Sulphuret of Antimony (which see.)

PROTO-CHLORIDE OF SULPHUR.—See Chloride of Sulphur.

SULPHIDE OF LEAD.—Occurs native as galena and is one of the ores of lead, having a specific gravity of 7.2 to 7.7. Commercially it is found as a black powder, of specific gravity 6.9. Its composition is 86.6 per cent. of lead and 36.4 per cent. of sulphur. Sulphide of Lead is a very useful black pigment, and one that is used quite largely in rubber works, as it is a good filler and assists in vulcanization. It is often made from pure white lead by very simple treatment. It materially assists the resiliency of Para compounds.

SULPHUR LOTUM.—A name for sublimed sulphur that has been washed to move sulphurous acids, and carefully dried.

SULPHIDE OF ZINC.—Sulphur forms with zinc two sulphides. One of these, the mono-sulphide, corresponds to zinc blende, which, as found native, is of various colors, from yellow to black. Its specific gravity is from 3.5 to 4.2. The other is a penta-sulphide artificially prepared and occurs in the form of a white powder. Upon ignition in the absence of air this latter substance loses four-fifths of its sulphur, but the temperature at which this takes place is too high to render it available as a source of sulphur of vulcanization in compounding rubber mixtures. With a slight addition of sulphur it is used in the production of white goods.

Sulphur occurs in a number of different forms, and under various names as brimstone, flowers or flour of sulphur, roll sulphur, rock sulphur, etc. Its specific gravity is 1.98 to 2.06. It melts at 239° F., thickens and becomes orange yellow at 320° F., at 428° it is semi-solid and red, and on carrying the heat higher it becomes browner and boils at 788° F. Some of the sulphur now used commercially is recovered from alkali waste, but most of it comes from Sicily, where it is found native. It is more generally used in rubber works than any other ingredient, and in all proportions from 3 per cent. up to 100 per cent. of the weight of the rubber. The ordinary form in which it is found in the rubber factory is in a yellow powder, known as flowers of sulphur. It has a slight affinity for moisture, and careful manufacturers keep it covered from air to avoid the formation of sulphurous or sulphuric acids. Mixed with certain oils by heat, it forms the black sulphur substitutes that are often used in rubber compounding. Sulphur in the form of rolled brimstone is pulverized, sifted, and used in the place of flowers of sulphur, in France, and is equally good and cheaper.

Sulphur Balsam.—A solution of sulphur in fixed oils, consisting of 2 ounces of flowers of sulphur in 8 ounces of linseed oil, used in proofing compounds.

Versuvian White.—A special vulcanizing material manufactured in England, for use in the manufacture of tennis balls and other goods.

Vulcanine.—An English vulcanizing preparation, used for both steam and dry heat goods. It occurs either as a white or a black powder, depending upon the line of goods on which it is to be used.

CHAPTER V.

FILLERS AND OTHER INGREDIENTS USED IN DRY MIXING IN RUBBER COMPOUNDS.

India-rubber is compounded for two reasons, the first being to reduce the cost without destroying the usefulness of the gum, the second being to impart to the gum qualities possessed by a great variety of mineral, vegetable, and even animal substances. Each of the ingredients treated in this chapter has some specific use. While their arrangement may seem a little incoherent to the chemist, it will be fully appreciated and understood by the rubber manufacturer whose habit of mind leads him to reach out into any of the kingdoms—animal, vegetable, or mineral—for assistants in compounding problems.

Acetate of Lead.—A white sweetish tasting powder soluble in water and alcohol. In its crystaline form it contains about 7 per cent. of water of crystalization, which is easily driven off at a temperature of say 80° to 100° F. Its specific gravity is: crystalized, 2.3; water free, 2.5. It is a product of the half completed process of treating pig lead where the old Dutch method of corrosion is employed in making the carbonate. Its use in semi-hard composition was patented by both Goodyear and Payen. India-rubber dissolved in oil, to which has been added acetate of lead, is used to fill the pores of certain leathers so that the "filling" shall not come through. It is also used in certain varnishes in connection with Gutta-percha.

Agalmatolite.—A silicate of aluminum and potassium resembling soapstone which is soft enough to be carved with a knife. It has no advantages over talc, silicate of magnesia, or soapstone in rubber use. It appears, however, in some patented compounds, but if the potash principle is necessary, it can be easily added to the ordinary powdered talc. The largest deposits of this material are to be found in China.

Antimony.—See Golden Sulphuret of Antimony, Black Antimony, and Kermes.

Argillaceous Red Shale.—The shales, clays, and feldspars are all very closely allied and pass the one into the other by gradual decay. A shale that has a large amount of clay in it is termed Argillaceous, and the substance mentioned in the heading may be briefly termed red clay tinctured with oxide of iron. The analysis of Argillaceous clay shows: Alumina 39, silica 46, water 13, iron, magnesia, and lime 2. It was the basis of a well-known oil resisting compound that for years baffled imitation.

ARTIFICIAL SULPHURET OF LEAD.—See Burnt Hypo.

ALUMINA.—The oxide of aluminum and a chief consituent of clay. Its specific gravity is 4.154. Ordinarily speaking it is a very inert substance, insoluble, and not readily attacked by acids. It is best known in the arts under the forms of kaolin, corundum, emery, etc. As obtained chemically it is a fine white glistening powder, feeling harsh and dry to the touch. Eaton's formula for the use of oxide of aluminum in making a pure white rubber, was India-rubber 40 per cent., oxide of aluminum 55 per cent., and sulphur 5 per cent. He describes his process for making the powder, which was by the burning of sulphate of alumina.

ANHYDRITE.—The water free mineral form of sulphate of lime or gypsum. It has a specific gravity of 2.9, and is formed artificially by heating gypsum so as to drive off all its water. It is white in color and crystaline in form. Gypsum that has been overheated in the preparation of plaster of paris and that has lost its ability to "set" is pure Anhydrite. It is used as a filler in rubber compounding instead of whiting or paris white.

ARSENIC.—A white brittle metal, with a specific gravity of 4.7 or 3.7, according to its form. Also a popular term for the oxide of arsenic sometimes called the white arsenic, which is a heavy white powder of the specific gravity 3.7. White arsenic is slightly soluble in cold water and to the extent of 10 per cent. in hot water. There are several coloring matters formed from arsenic, all of which are to be condemned for general use. The most familar are paris green, realgar, which is red, and orpiment, which is yellow. The white oxide is rarely used in rubber work, and is to be avoided, as are the greens, reds, and yellows. The green has been used in mechanical rubber goods, but the color was not a valuable one. Hancock vulcanized Gutta-percha with orpiment, and Forster used it in "mosaic work" for floor coverings. An anti-fouling composition for ships' bottoms is formed of Gutta-percha, copper, bronze, and arsenic. Another is formed of India-rubber 2 pounds, rosin 7 pounds, and arsenic 2 ounces.

ASBESTIC.—The part of the rock remaining after the richer veins of asbestos have been extracted. This remainder is a purely fibrous material, clearly showing its origin. For mechanical uses it is ground fine, and for all sorts of fireproofing purposes is valuable and much cheaper than long fiber asbestos. It is mined at Danville, Lower Canada. It makes an excellent compounding material for asbestos packings, etc., in connection with rubber.

ASBESTINE.—A pure fibrous silicate of magnesia, called also mineral pulp. It is mined near Gouverneur, N. Y., where is the only deposit at present known where magnesia shows so distinct a fiber. It is very largely used in the manufacture of paper, and also as an ingredient in rubber. Apparently the pulverized mineral is a very strong white powder, but in actual use it has not much more covering quality than whiting. It was at one time used largely in the manufacture of rubber shoes, but, aside from being inert and a good filler, was probably no better than whiting, while it was more costly. It is often used in white goods, in connection with oxide of zinc to make a light weight compound. It is also known as agalite and asbestine pulp. Its composition is: Silica 62, magnesia 33, water 4, iron oxide and alumina 1.

ASBESTOS *(Amianthus)*.—A fibrous silicate of calcium and magnesia, also called stone flax, Salamanda's wool (from an old belief that it was originally made from the wool of the salamanda), cotton stone, mountain flax, mountain wood, and mountain cork. Its specific gravity is 3.02 to 3.1. An analysis of the 2 best known varieties shows:

	Canadian.	Italian.
Silica	40.92	40.25
Magnesia	33.21	40.18
Water of hydration	12.22	14.02
Alumina	6.69	2.82
Protoxide of iron	5.77	.75
Soda	.68	1.37
Potash, etc	.22	.15
Sulphuric acid	traees	.31

The longest fiber is possessed by the Italian, which is sometimes 3 feet in length. The Canadian ranges from 3 to 6 inches in length, but it is finer, more flexible, and more easily separated than the Italian. The mineral divides iteslf naturally into 3 classes: The first, coarse, brittle, very plentiful, and cheap;—the second, possessing well-defined fibers of a brownish yellow color, fragile, and containing many foreign bodies; the third, with pure white silky fibers which can be woven into textiles. A notable use to which asbestos has been put in United States is in the production of the packing known as Vulcabeston (which see). Its low conductivity of heat renders it particularly useful in steam packings, both for cylinder work and for joints, while its incombustibility has long caused it to be used for fireproof purposes. There are fibers formed of serpentine rock which are much used as a substitute for genuine asbestos, and answer nearly as well, being, however, shorter in fiber and somewhat less durable. Almost all large rubber manufacturers produce packings in which there is a certain amount of asbestos, often assisted by infusorial earth, asbestine, etc.

ATMOID.—A very light white earthy matter, marketed by an English corporation. Analysis proves it to be an almost pure silica —quite close, in fact, to infusorial earth.

BARYTES.—A heavy white mineral that in commerce takes the form of a fine white or gray powder. It is obtained by grinding the mineral heavy spar, or by chemical means from baric chloride. Its specific gravity is 4.5. It occurs in commerce under the names "permanent white" and "blanc fixe." The artificially prepared substance is to be preferred to the finely ground mineral, on account of its less crystaline form. The commercial article should always be examined to determine its freedom from acid impurities. Barytes is also called Witherite, which is the carbonate, and Heavy Spar, which is the sulphate. Barytes is chiefly used as an adulterant for white lead and paints. Thus Venice white contains equal parts of sulphate of barytes and white lead; Hamburg white, 2 parts to 2 parts of white lead; and Dutch white, 3 parts to 1 of white lead. It is wholly inert when used as an ingredient in rubber compounding, and increases the resiliency of rubber, and is a make-weight.

BLACK ANTIMONY.—A black powder obtained by grinding stibnite or antimony ore. It is a sulphide of the metal and is met with more or less pure, as it is often prepared from a high grade ore. The sulphur contained in it is unavailable for vulcanizing purposes, and if used in compounding it is necessary to add a sufficiency of sulphur to vulcanize. In the purest form black antimony contains about 28 per cent. of sulphur and 72 per cent. of antimony. It is insoluble in water, but is dissolved by muriatic acid or by caustic alkalies. From its solution in alkali a fine brown red powder may be obtained by treatment with a dilute acid, and this powder, known as kermes, has the same chemical composition as that mentioned above. Its specific gravity is 4.6. It was formerly used sometimes as a filler, as it was believed to give a soft effect in molded goods. It has been almost wholly displaced, however, by cheaper and better ingredients.

BLACK HYPO.—See Hypo-sulphite of Lead.

BLACK LEAD.—See Plumbago.

BLUE LEAD.—Where zinc ores are found in combination with galena, or natural sulphide of lead, the two are often smelted together with raw coal and slaked lime, producing a fume called blue powder, which is sold under the name of Blue Lead. It is an excellent filler, but is not as good as sublimed lead, for example, as it does not impart enough resiliency to rubber. Its chief merit is its cheapness. A very fine quality of Blue Lead, containing considerable lead oxide, is now on the market, but this must not be confused with either of the two low grade articles mentioned in these paragraphs. This Blue Lead is of exceeding fineness, and gives a peculiarly soft finish to the rubber. Used in the place of litharge, it materially assists in the cure, and produces a fine black. As it has a high specific gravity, it often displaces barytes. Blue Lead is also a name given to an artificial aluminous substance occurring either as a loose powder or in a concrete form, colored blue by means of some kind of blue dye—aniline or logwood—which does not contain lead.

BONE ASH.—See Phosphate of Lime.

BONEBLACK.—See Animal Charcoal.

BUCARAMANGUINA.—A transparent amber colored, incombustible material, found near Bucaramanga, Colombia. It is somewhat similar to asbestos, for which it has been mentioned as a substitute in the manufacture of packings.

BURNT UMBER.—An earth containing a large amount of iron

oxide of a dark brown rust color. As mined it is called raw umber, and the product obtained by calcining it is known as Burnt Umber. It is a fairly useful filler in compounding, as its action, or rather lack of action, upon rubber makes it safe to use. It is used in brown packings and, to a certain extent, in maroon goods.

CALAMINE.—An ore of the metal zinc, and a carbonate of zinc. Ordinary Calamine, which is a silicate of the metal, has a specific gravity of 3.6 to 4.4, and is little used in the arts. Noble Calamine, or native carbonate of zinc, is a gray or grayish yellow to brown powder, according to its priority. Its specific gravity is 3.4 to 4.4. Its nature is earthy, and heat has no action upon it. A little of it is said to toughen soft compounds.

CALCIUM WHITE.—Another name for Whiting.

CALOMEL.—A white, tasteless, and inodorous powder of specific gravity about 7.2. It is permanent in the air, but should be kept in the dark, as light blackens it. When pure it may be wholly volatilized by heat, but if this cannot be done, then the sample tested contains other bodies. Calomel strikes a black color under the action of alkalies. It is insoluble in water, alcohol, ether, or benzine. It is the basis of a compound for rendering woven hose waterproof, the other ingredients being magnesia, black antimony, oxide of zinc, tar, sulphur, and India-rubber. Its office is to hasten the cure.

CARBONATE OF BARYTA.—Known also as the mineral witherite; has a specific gravity of 4.3. It is a white powder insoluble in water and alcohol. (See Barytes.)

CARBONATE OF LEAD.—See White Lead.

CARBONATE OF LIME.—Very familiar under the form of limestone, marble, or chalk. Specific gravity 2.7 and 2.9. (See Whiting.)

CARBURET OF IRON.—A name given to a mixture of graphite and oxide of iron. A fine black-brown powder, fairly heavy specifically, although variable. It makes a fair filler in compounding being inert and strongly coherent. In packings it has been largely used and also in compounds for wagon covers and tarpaulins before reclaimed rubber came largely into use. It has also been used in cements for card clothing.

CHALK.—A white soft, somewhat gritty substance, consisting chiefly of carbonate of lime. It is made up of myriads of very small shells of marine animals long extinct. Its nature is earthy; that is to say, it is not easily affected by ordinary bodies. Acids disengage carbonic acid gas from it. Its specific gravity is 2.9. If heated to a red heat, carbonic acid gas escapes and quicklime is left behind. (See Whiting.)

CHARCOAL (ANIMAL).—Animal charcoal is made from calcined bones and has the property, in a high degree, of absorbing odors. It is often used, therefore, in deodorizing rubber goods, and experimentally by chemists for filtering Gutta-percha dissolved in bisulphide of carbon, where a perfectly clear product is desired. Its use is advised by Forster in Gutta-percha compounds, and by Warne, Jaques, and others for making packings to stand a high degree of heat. (See Boneblack.)

CHARCOAL (VEGETABLE).—This is a popular term for the coal produced by the charring of wood. There are many materials which are really charcoals, such as animal charcoal just quoted, carbon, coke, graphite, and wood charcoal. All of these are practically the same in their pure states, being almost wholly carbon. Wood charcoal, which is what is meant in rubber compounding by vegetable charcoal, consists of carbon, hydrogen, and oxygen, the last two being in the proportion to form water. As it retains the form of the wood from which it is made, it is powdered before use. It is black and brittle, insoluble in water, infusible, and non-volatile in the most intense heat. It has the power of condensing gases and destroying bad smells. Charcoal may or may not be a bad conductor of heat and a good conductor of electricity, these properties depending upon the wood from which it is made. Technically, it is divided into hard wood charcoal and soft wood charcoal. Its composition at ordinary temperatures is about as follows: Carbon 85 per cent., water 12 per cent., ash 3 per cent. It is used in rubber compounding in certain vulcanite varnishes and in certain insulated wire compounds. For this latter use, willow charcoal is preferable, as it is a decided non-conductor. It has also been used in sponge rubber, with the idea that it acts as a preservative in a compound which is very likely to be short lived. One curious use for it, a possible and valuable one, was in the attempted manufacture of cop tubes from Gutta-percha and Charcoal. Macintosh also used large quantities of ground charcoal in place of lamp-black in some of his compounds. A French substitute for vulcanite paints or lacquers is made of 10 pounds of bitumen, 15 parts of Charcoal, and a little linseed oil, mixed by heating.

CHINA CLAY.— See Kaolin.

COMPO.—A name for a composition used in rubber manufacture in the United States years ago, but not in use now. The name, however, clings to two compounds sold by an English chemical house for use in rubber work. They are of a secret nature. No. 1 is used in the manufacture of oil-resisting valves and in tubing for chemical factories, in the proportion of 30 pounds of Compo to 10 pounds of rubber. No. 2 is used for soles for tennis shoes and in mechanical goods, in the proportion of 25 pounds of Compo to 10 pounds of rubber.

CORNWALL CLAY.—See Kaolin.

CORK, in granulated or powdered form, has long been a favorite ingredient in rubber compounding. Not that it is used in any such measure as whiting or barytes, but many mills have used it, and a few in large proportions. Used in connection with India-rubber and Gutta-percha, it has been the subject of some fifty patents. Its largest use, perhaps, was in the manufacture of Kamptulicon, where India-rubber is used as a binding material, and in linoleum, where oxidized oils are used in place of rubber. It was also used in what was known as leather rubber, in which palm oil distillate, a little India-rubber, and a good deal of granulated cork were used. At one time it was also compounded with rubber and made up into a waterproof felt for hats. It also went into compounds to resist heat, into cricket balls, and into golf balls, where it was compounded with Gutta-percha and enough metal filings added to give the necessary weight. A rubber blanket used in special manufacture also had its surface covered with granulated Cork as an absorbent material. In some cases the Cork was charred and roasted to remove what resinous matter might be in it, while in others resinous matter was removed by boiling in alcohol. As is generally known, Cork is the bark of the cork oak, a native of the south of Europe and north of Africa. The chief supplies come from Spain and Portugal. Cork is the basis of the fine black known as Spanish black, which is made by burning the refuse in close vessels.

CORUNDUM.—A mineral which is nearly pure alumina, yet of great specific gravity, and of exceeding hardness, being inferior, in this respect, only to the diamond. Emery (which see), so largely used as a polishing substance, is a variety of Corundum.

DIATOMACEOUS EARTH.—See Infusorial Earth.

ELECTRIC FACING.—See Farina.

EMERY.—The average composition of Emery may be taken as alumina 82, oxide of iron 10, silica 6, lime 1½. Its specific gravity is about 3.8 to 4. It is prepared by breaking the stone at first into lumps about the size of a hen's egg, then running it through stamps, and crushing it to powder. It is then sifted to various degrees of fineness, and graded according to the meshes of the sieve. Emery is next in hardness to diamond dust and crystaline corundum, and it is used chiefly as an abrading agent. Prior to the invention of vulcanite, emery wheels were made by mixing clay and emery in suitable mounds, and vitrifying them like common earthenware. In rubber mills it is chiefly used in the manufacture of what are known as vulcanite emery wheels. It is also used in grinding and sharpening compounds, as hones and strops. (See also Alumina and Corundum.) A certain amount of it also gives the desired surface to rubber blackboards.

FARINA.—This is sometimes used in small quantities in unusual mixtures as a compound, but has little value, as there are many better substitutes for it. A practical use for it, however, is the brushing of a rubber surface with it before vulcanization, when it is necessary to have printing or stamping done upon that surface afterwards. Farina is made largely of potatoes, another name for it being Potato Starch. The process consists simply of crushing, sifting, washing, bleaching, and grinding, which is repeated three times, and each time the starch granules separate and are collected. Potato Starch will be remembered by rubber manufacturers as the material which the gossamer makers used successfully for a number of years in the production of the "electric" or "corruscus" finish. Bone ash is used sometimes in the place of Farina, where rubber surfaces are to be printed upon.

FELDSPAR.—A name given to a group of silicates of which

the principal ones are Orthoclase or potash, containing silica, alumina, and potash, and having a specific gravity of 2.5; Albite, containing silica, alumina, and soda, specific gravity 2.61; Oligoclase, containing silica, alumina, soda, and lime, specific gravity 2.66; and Anorthite, containing silica, alumina, and lime, with a specific gravity of 2.75. The feldspars by the action of the weather break down into china clay, kaolin, or pottery clays. Ground very fine, they have been used in the production of rubber enamels and lacquers.

FIRE CLAY.—A kind of clay which, better than any other, resists the action of heat and direct flame. It is composed principally of silica and alumina, with traces of the alkali earths. The best is found in conjunction with coal, and is called Stourbridge clay. Its specific gravity it about 2.5, and its color dirty white. Mixed with vulcanized India-rubber, dissolved in tar oil and sulphur, it forms a compound which, when applied to hot joints, cures at once.

FLINT is practically pure silica and has the specific gravity of 2.63. The nature of the powder obtained by grinding is always sharp and gritty. It is unacted upon by all ordinary means, and with difficulty even in the laboratory of the chemist. Its principal use, perhaps, is in the manufacture of glass. Flint varies in color from yellow and brown to black. It has been used in erasive rubbers, although pumice stone is better.

FLOUR OF GLASS.—Glass powdered and sifted through a fine sieve of 150 meshes to the inch. Glass varies much in its composition, the more common kinds containing lime, while the so-called flint glass contains lead. Potash and soda also enter into the composition of glass; hence all flour of glass will contain those ingredients which entered into the composition of the glass it was obtained from. Generally speaking, Flour of Glass may be considered an inert substance under ordinary conditions, though the softer kinds are attacked even by boiling water. It was used by Newton and Wray in insulated wire compounds, and has also been used in certain packings.

FLOUR OF PHOSPHATE.—See Phosphate of Lime.

FOSSIL FARINA, also called mountain milk, is an earth similar to infusorial earth. It is obtained from China and consists of silica 50½, alumina 26½, magnesia 9, water and organic matter 13, with traces of lime and oxide of iron. It has been used in rubber compounding for the production of packings and semi-hard valves.

FOSSIL MEAL.—A kind of earthy mineral, principally composed of the minute shells of very small animals long extinct. It is similar to infusorial earth, lime and silica entering chiefly into its composition. It is used for the same purposes as infusorial earth (which see) or silica.

FRENCH CHALK.—This is ground and sifted talc, forming a white, greasy-feeling powder. Its chemical composition is hydrated silicate of magnesia, the water being chemically combined. Its specific gravity is 2. (See Talc.)

FULLER'S EARTH.—A kind of clay. It is a greenish or brownish earthy, somewhat greasy-feeling, substance, having a shining streak when rubbed. Its composition is: Silica 70, oxide iron 2.5, alumina 3.5, lime 6, combined water 16, magnesia trace, phosphoric acid trace, salt 2, alkalies trace. Fuller's Earth is found in extensive deposits in England, where its annual consumption at one time exceeded 2,000 tons, chiefly in the woolen manufacture, for fulling cloth. Its specific gravity is from 1.8 to 2.2. It is used in rubber compounding for about the same purposes as infusorial earth, and is also used in the manufacture of rubber type.

GRAPHITE.—See Plumbago.

GYPSUM.—See Sulphate of Lime.

INFUSORIAL EARTH.—This is obtained usually from deposits at the bottom of inland waters, and consists of the minute siliceous remains of infusoria or microscopical animals. It is known also as fossil flour, mountain flour, and infusorial flour. The largest deposits, in the form of a fine white or pinkish powder, are found in Nova Scotia and in Germany. This earth is a wonderful nonconductor of heat, and, in connection with asbestos, is used in the manufacture of boiler coverings. It is used also in small proportions in various rubber compounds, where it increases both strength and resiliency, though if used in excess it makes a very hard compound. The best grades are wholly free from vegetable matter, are nearly pure silica, and perfectly indifferent to corrosive substances. Under the name of diatomaceous silica it is used in a formula for elastic valve packing, patented by A. B. Jenkins, United States. This packing is described as practically indestructible in steam or water, oils, acids, etc.

Iron Pyrites.—A sulphuret of iron, commonly of a bright, brass yellow color; a very plentiful mineral often mistaken for gold. It is used in the manufacture of sulphuric acid, while sulphur is also obtained from it by sublimation. It was used by Warne, Fanshaw, and others in the manufacture of packings to resist a high degree of heat. The sulphur in Iron Pyrites has also been used in vulcanization. Warne, in one of his heat resisting packings, patented the use of Iron Pyrites, and, in the compound that he gives as an example, leaves out the whole or a portion of the sulphur usually employed. (See Vulcanization.)

Kermes.—A brownish red form of sulphide of antimony, artificially prepared by boiling in carbonate of soda. If left to itself the solution will partly deposit a very fine powder of Kermes, while the clear solution may be further treated with a weak acid to obtain the remainder. Kermes will not vulcanize rubber without the addition of sulphur. Its specific gravity is about 4.5. Its composition is 28 per cent. sulphur and 72 per cent. antimony. It is rarely used in rubber compounding.

Lime.—The oxide of the metal calcium. It is commonly known in two states, viz.: Quick Lime, which is the pure oxide, and Slaked Lime, which is the hydrated oxide mixed with some carbonate. Quick lime is a white solid substance of specific gravity 3.2. It is not stable, taking up water and carbonic acid from the air and breaking down into a fine white powder, usually called air-slaked lime. Its power of absorbing water has caused it to be favorably used in drying operations, while the insoluble compounds it forms with various oils have led to its being considered as a drier, although this action is not properly to be called one of drying. Lime air slaked is used in rubber work, where there may be a little moisture in a compound, which it readily neutralizes. It is also used in soft cements in connection with tallow and India-rubber, but only where the rubber has been melted and the cement is of the non-drying variety. In compositions like that of Sorel's, Lime is introduced to effect a combination between resin acids found in the resin and resin oil. Excess of Lime in India-rubber is injurious, because it renders the compound too open, thus inducing oxidation. When used in small quantities, aside from its effect upon moisture, it combines with free sulphur and modifies its continued action upon the rubber. It must be remembered, however, that lime diminishes the resiliency of India-rubber, while it increases the hardness of both hard and soft rubber. It may be used in small quantities in insulated wire, and in a measure assists the insulating capacity of the rubber. Calcium carbonate, in connection with colcothar and methol alcohol, is used as a compound for cleansing vulcanite. Rubber also cures quicker when compounded with Lime.

LITHARGE.—One of the oxides of the lead, known as the monoxide. When pure its specific gravity is 9.36. Commercial litharge often contains carbonic acid gas and water taken up from the air. These may be removed by strong heating. It has a peculiar property, the nature of which is yet a debated question, by virtue of which it renders oil more easily oxidized, or, as it is commonly called, rendered dry. There is no reason to suppose that this action is available with caoutchouc. The best Litharge is made from pig lead, which is placed in a reverberatory furnace and exposed to a current of air, which reduces it to an oxide. It has been noted in rubber factories that certain men seem specially sensitive to the effects of Litharge, often developing serious symptoms of lead poisoning. Persons who show any symptons should pay scrupulous attention to personal cleanliness. It is said that such persons have been cured by taking them out of the mixing room entirely, and putting them to work on vulcanizers, particularly where they open and handle the goods from the finished heat, the theory being that the sulphur fumes neutralize the effects of the leads. Possibly there is a grain of wisdom in this, for the old fashioned treatment for lead poisoning was sulphur baths and the drinking of water acidulated with sulphuric acid or the acid or sulphate of magnesia. Litharge is not only a valuable filler for rubber, but has the faculty of hastening vulcanization in a marked degree. All dry heat goods depend upon it, and in mold work and general mechanical goods it is used whenever possible. Of course, it is generally available for dark or black effects only.

LITHOPHONE.—See Colors.

LITHARGITE.—A substitute for litharge, made of a mixture of pulverized and calcined magnesia and oxide of lead.

MAGNESIA.—The oxide of the metal magnesium. A white dry powder which, when mixed with water, forms a hard compact mass like marble. Its specific gravity is 3.65. It is earthy in its nature, having no taste, but producing a sense of dryness in the mouth owing to its absorption of the water therein. It is frequently called calcined magnesia from the method of preparation by burning magnesia alba. Its use in rubber is to increase its toughness and resiliency, which it does to a marked degree when used in moderation. Magnesia is also used in the production of compounds like balenite, its use in hard rubber compounds being to increase resiliency as well as hardness. A very small quantity of it is also used in compounds for insulated wire, where it is said to increase the insulating qualities of rubber. Carbonate of magnesia occurs native in the mineral magnesite and, in connection with carbonate of lime, as dolomite.

MANGANESE.—A metal of the iron group; gray or reddish white in color, and must be kept under rock oil or in well sealed vessels, being easily destroyed by the air. Its specific gravity is 7.2. Manganese is obtained artificially as a black powder, by exposing the peroxide to prolonged heat. When ignited it is converted into a red oxide, which corresponds to the black oxide of iron. The black Manganese of commerce is the peroxide. Oxides of Manganese have a destructive effect on rubber and blacks that contain this, as they sometimes do, are to be avoided. Manganese is used in connection with pitch, turpentine, and Gutta-percha for making Brandt's cement.

MARBLE FLOUR.—This is the finely ground chips of white marble, and is composed almost wholly of carbonate of lime. It is a heavy inert powder, often used in rubber compounding as a susbtitute for barytes. It has also been used to some extent in hard rubber, and in the manufacture of hones.

MASSISOT.—An oxide of lead, dull red orange in color. A higher degree of oxidation turns this into a product called Minium, which is its purest state. It is often used in rubber compounds, acting practically like litharge.

MICA is the name given to a group of complex silicates containing aluminum and potassium, generally with magnesium but rarely with lime. Their specific gravity ranges from 2.8 to 3.2, while their color varies greatly. Ground mica is simply one or other of these micas reduced to powder. It is used in rubber compounding chiefly for insulating purposes. It is handled as a cement, compounded with rubber, and cut with benzine, or may be mixed dry on the grinder. It is also used in fireproof coverings in connection with rubber, and it is said that for a semi-hard result that is to come in contact with hot water, rubber and Mica forms the best compound. Mica in a state of a very fine powder is also known as "cat's gold" or "cat's silver."

Mineral Wool.—Produced by sending blasts of steam through molten slag, which reduces the fluid metal to a fiber similar to the fused glass that is spun into glass silk. Natural mineral wool, such as is found in the Hawaiian Islands, is very brittle, but the artificial has considerable toughness. It is also known as slag wool, or silicate cotton. It appears in light fleecy masses, and at a distance looks like fine cotton batting. It is very cheap, but is easily affected by weak acids, and should be kept away from a moist atmosphere. It has not been largely used in rubber work as yet, but Lascelles-Scott strongly advises its use, giving as reasons its cheapness and its physical fitness. The sulphides present in it also assist in vulcanization.

Minium.—One of the oxides of lead, known also as Red Lead (which see). It is a scarlet crystaline and granular powder, having a specific gravity of 8.6 to 9.1. On heating, it temporarily changes color to violet and black, but returns again to the scarlet on cooling. It is adulterated with oxide of iron and brick dust.

Mountain Flour.—See Infusorial Earth.

Orange Mineral.—A red lead made from carbonate of lead, while red lead is made from litharge. As a general rule, it contains some lead carbonate. It differs from red lead in color, in that it is more orange red, and more brilliant. The reason for this difference is that it is less crystaline, its particles being much finer than those of red lead. The pigment is also more bulky and much smoother. It is used in finer grades of dark rubber, to assist the cure and impart resiliency.

Oxide of Aluminum.—See Alumina.

Oxide of Antimony.—There are really three of these oxides. The tri-oxide, one most useful in the arts, is a snow white powder of the specific gravity of 5.2. It may be obtained by treating stibnite or, better still, powdered antimony metal with nitric acid, in a current of air sufficient to carry off the copious fumes arising during the operation, or by treating the chloride of antimony with cold water for several days. A mixture of the tri-oxide with a small percentage of the insoluble peroxide may be obtained by melting antimony in a cast iron retort fitted with nozzles, through which air may be blown so as to bubble through the melted metal. Dense white fumes arise, which may be condensed in suitable chambers into a snow white powder. This is used in coloring dental vulcanite.

Oxide of Gold.—As a matter of curiosity it may be noted that this is the most costly ingredient suggested for rubber compounding. It occurs in two forms—the protoxide, a dark green or bluish violet powder, and the teroxide, a brown powder. The use of the protoxide was patented by Ninck. For dental vulcanite is is doubtful if either form of the oxide could be used, even if the price were so low as to bring it within reach. Another formula calls for the mechanical admixture of gold leaf, which is practicable—if one possesses the gold.

Oxide of Lead.—See Minium and Litharge.

Oxide of Tin.—The article most frequently used in the arts is the di-oxide. This is a white water-free powder, of the specific gravity of 6.7, insoluble in acids and such solvents as naphtha, petroleum, etc. It is infusible, except at a very high temperature, and is tasteless and inodorous. What is known as French Oxide of Tin is simply a carefully prepared and purified form of the di-oxide. It is rarely used in rubber work, although Newton recommends it for a basic ingredient in rubber type. The other oxides of tin are at present merely of chemical interest.

Oxide of Zinc.—See Colors.

Oxychloride of Lead.—There are several oxychlorides of lead. The substance once known as Turner's Yellow and another known as Carsel Yellow were both of this composition. More recently a white compound has been prepared, which, from its covering power, has been used largely as a paint. Tarpaulin compounds consisting of India-rubber, coal tar, and pitch are treated with Oxychloride of Lead for surface drying, in lieu of vulcanization.

PAGODITE.—A mineral resembling steatite or soapstone. Its name comes from its having been used in the East as a material for carving miniature temples or pagodas from, as it is soft enough to be cut with a knife. Its specific gravity is about the same as that of soapstone, and its color greenish white. (See Agalmatolite.)

PARIS WHITE.—This has exactly the same composition as Whiting, but is a much harder and more compact form of English chalk, and therefore has greater density. Spanish White is a coarser variety of the same material. Its uses are practically the same as those of whiting.

PETRIFITE.—A white powder composed of two inexpensive but secret substances. When mixed with water it solidifies quickly, and is an excellent binding substance. Mixed with marble dust, it is sometimes melted and cast upon glass or other smooth surfaces, and makes an excellent table top in place of the zinc tables used in many rubber factories. As it is perfectly impervious to ordinary solvents, neither cement nor India-rubber sticks to it. It is manufactured in England.

PEROXIDE OF LEAD.—The highest oxide of lead—a dark brown powder with a specific gravity of about 9. It is easily decomposed, and from this characteristic it has a strong oxidizing action. Exposed to sun light or to heat, it yields oxygen and passes into the lower oxide known as Red Lead. Its oxidizing properties make it a questionable ingredient in compounding rubber, although certain formulas call for its presence.

PEROXIDE OF MANGANESE.—Another name for Black Oxide of Manganese, which is a black powder having a specific gravity of 4.8. It is not readily acted on in ordinary ways, being unchanged by heat short of bright red. It is insoluble in the ordinary hydrocarbon solvents. Solvent naphtha was treated with Peroxide of Manganese by Humphry to free it from water. (See Manganese.)

PHOSPHATE OF LIME.—The chief constituent of animal bones, forming the bulk of the ashes of the same when burnt. It is a white powder, and when in crystaline mineral form, it has a specific gravity of 3.18. It is insoluble in ether, alcohol, or the benzine class of solvents. As it occurs naturally it is known as flour of phosphate and is used in part as a substitute for whiting. Bone ash made from animal charcoal is used in the same way.

PHOSPHORUS.—A non-metallic element or metalloid, although in its combining relation it is more closely connected with arsenic and antimony than with any members of the sulphur group. It is found ordinarily in two states—the ordinary phosphorus and the red variety. Ordinary phosphorus is an almost colorless or faintly yellow solid substance, somewhat resembling wax, and giving off a disagreeable odor. It fuses at 111.5° F. into a colorless fluid. Heated in the air to about 140° F., it catches fire and burns with a bright white flame. It dissolves freely in benzol, bisulphide of carbon, and in many oils. Red phosphorus is an amorphous powder of a deep red color, with no odor, and may be heated to nearly 500° F. without fusing. Its specific gravity is 2.10. It does not take fire when rubbed, undergoes no change on exposure to the air at ordinary temperatures, and is far less inflammable than ordinary Phosphorus. It is insoluble in solvents of the ordinary Phosphorus, and is not poisonous. Mulholland made an insulated wire compound from shellac and India-rubber in solution, combined with 1 to 2 per cent. of Phosphorus, which he cured with chloride of sulphur. As cold-cure gums are of little value as insulators, his invention is of doubtful value. He also made a prepartion of India-rubber, resin and tallow, and shoddy, to be applied in a fluid state where gas came in contact with the rubber, adding Phosphorus after his solution was finished, to prevent decomposition of the rubber. Duvivier also treated Gutta-percha with sulphide of phosphorus, claiming that he got an elastic result, but allowing that his compound was damaged by acid vapors, to neutralize which action he mixed carbonate of soda with it. An anti-fouling preparation of English origin was also made of Gutta-percha, turpentine, and a little Phosphorus.

PIPE CLAY.—A peculiar kind of clay containing neither iron, sand, nor carbonate of lime. It is a beautiful white, retaining its whiteness when burnt. It belongs to the group of clays. Its specific gravity is 2 to 2.5. It was used by Mayall in combination with Gutta-percha, India-rubber, zinc, shellac, and resin for insulating tape, and by Day to absorb gases during vulcanization.

Plaster of Paris.—This is prepared from gypsum or sulphate of lime. Its properties of hardening when made into a paste with water are well-known. Its chemical properties are the same as burnt gypsum. It is used sometimes instead of lime in compounding and also for making trial molds for rubber work. It was used in old fashioned dry heat compounds to prevent blistering. (See Anhydrite.)

Plumbagine.—A dark colored pigment manufactured in England and sold to rubber manufacturers for the production of valves. By its use the rubber is vulcanized and goods made which are said to resist successfully the action of cheap lubricants. One pound of Plumbagine is used to 2 pounds of rubber.

Plumbago.—This sometimes is called Black Lead, though having no relation to lead; it is also called Graphite. Its specific gravity is 2.1 to 2.2. Its color is black and shiny. It consists chiefly of carbon, but contains more or less alumina, silica, lime, iron, etc. varying from 1 to 47 per cent., but not chemically combined. Black Lead is a perfect conductor of electricity. It is more incombustible than most ingredients used in rubber compounding, and is capable of withstanding great heat. It is used in the rubber industry, chiefly in the manufacture of what are known as graphite or plumbago packings. It is a wholly inert substance, safe to use in connection with any compounds, and is not affected by heat or acids, alkalies, or corrosive substances. It is useful also in certain polishing compositions made with India-rubber as a base. German asbestos cements almost all contain a good proportion of finely powdered graphite.

Portland Cement is obtained by burning the mud found at the mouths of several large rivers in Europe with a proportion of clay and lime. Its composition is somewhat complex, containing: Lime 55 to 63 per cent., silica acid 23 to 26 per cent., alumina 5 to 9 per cent., and oxide of iron 2 to 6 per cent., together with magnesia, potash, soda, sulphate of lime, clay, or sand in various small proportions, according to the mode of manufacture. Its value as a cement depends upon the interaction of the lime and the silicic acid. In compounding it would have no chemical effects upon rubber, but might of itself become much hardened and thus cause mechanical injury to goods in which it has been introduced. As it occurs commercially, it is a gritty powder of a gray brown or yellow brown color. The gray brown makes the best cement. Its only use as far as known in rubber is where it is mixed with tar oil and waste rubber to joint pipes containing fluids.

POWDERED COAL.—Coal consists chiefly of carbon, and is universally regarded as being of vegetable origin. Various coals differ widely in their composition and characters, running from the softest kinds of earths to compact and solid bodies like Parrot coal, which is so compact and solid that it has been made into boxes, inkstands, and other articles which resemble jet. The average specimen of coal analyses is: Carbon 82.6, hydrogen 5.6, oxygen 11.8. Some curious compounds of India-rubber and Coal have been formed. One, for instance, was a mixture in which 2 pounds of waste India-rubber in a cheap solvent was mixed with nearly a ton of powdered Coal, in which was a certain amount of clay and peat, the use being for an artificial fuel; another use was in the production of hard rubber. Indeed, it is probable that the cheapest compound in use to-day is a jet black, semi-hard rubber made almost wholly of powdered bituminous Coal in which is incorporated a very small percentage of rubber. Coal that is to be used in any rubber work should be submitted to a chemist and its sulphur and other compounds carefully determined before use.

PUMICE STONE.—A light porous ashy stone, the product of volcanic action, its structure being that of a mass of porous glass. Its composition is a mixture of silicates of aluminum, magnesia, calcium, iron, potassium, and sodium, varying with the particular lava whence it had its origin. Its action on India-rubber will be quite inappreciable, chemically speaking, but its mechanical action will be that of a sharp cutting powder. Ground fine, it is used in the manufacture of erasive rubber, and is also used compounded with the rubber in the manufacture of hones. Recent patents call for its use in certain semi-hard compounds, its presence being said greatly to increase the toughness of the compound. Mixed with lard oil to a thick paste, this has been used for polishing India-rubber.

PUZZOLANA.—A porous lava found near Naples, used chiefly, when mixed with ordinary lime, forming hydraulic cement. Compounded with marine glue, it is used as a varnish for preserving metallic articles from corrosion.

RED CHALK.—Artificially deposited chalk colored by any suitable pigment—usually one of the red oxides of iron. (See Chalk.)

RED LEAD.—An oxide of the metal, which is also known as Minium. Prepared from pure massicot or from white lead. Its specific gravity is 8.6 to 9.1. A scarlet crystaline granular powder, of rather strong coloring powers. As a colorant in rubber work it would be unavailable, since the sulphur necessary to vulcanize would render it more or less black, owing to the formation of sulphide of lead. It is sometimes used, however, in place of litharge. It is also used in "hot" cements of Gutta-percha and for varnishes such as those made of India-rubber, linseed oil, etc., for covering the backs of mirrors. (See Minium, Massicot, and Orange Mineral.)

ROTTEN STONE.—Usually considered to be the residuum of naturally decomposed impure limestone, and varying in composition with its sources. That from Derbyshire, England, shows much alumina; other sorts have more silica. The name is sometimes given to "tripoli" which is a species of infusorial earth. It can have no particular action on rubber, as it is very inert, but is used in certain packings, and was also used by Warne in insulated wire compounds.

SELENIUM.—A non-metallic element or metalloid of a dark brown color, analagous to sulphur. It has no smell, is tasteless, and is a non-conductor of electricty. It occurs rarely in nature, being found chiefly as a selenide in combination with lead, silver, copper, or iron. It is the basis of a process for vulcanizing India-rubber.

SILEX.—Pure silica. (See Flint.)

SILICA.—The oxide of the metal silicon, familiar in the forms of flint, quartz, etc. Its specific gravity is 2.6. It is without action on India-rubber, except mechanically speaking. It is used in Chapman's vulcanite enameling solution, made of Inida-rubber, sulphur, and Silica. (See Flint.)

SILICATE COTTON.—See Mineral Wool.

SLAG WOOL.—See Mineral Wool.

SLAKED LIME.—Quick lime that has been treated with water, and allowed to absorb it from the air and crumbled to a fine powder. (See Lime.)

SLATE.—A soft easily laminated earthy material, chiefly aluminuous in composition, and allied to the clays. Finely ground, it makes a good semi-hard valve of a blue gray shade. It has been also used in general rubber compounding.

SOAPSTONE.—A silicate of magnesia, combined with more or less alumina and water. It is really a massive form of talc. In color it is white, reddish, white, or yellow, is soft and greasy to the touch, is easily cut, but is hard to break. Its specific gravity is 2.26. It is used often in the place of French talc, for keeping rubber surfaces from sticking together during vulcanization, and also for burying dark colored goods and holding them in shape while they are being cured. Used as an adulterant for rubber, it makes an excellent semi-hard compound for valves. It is also used as a basis compound in the manufacture of insulated wire. (See Talc.)

STARCH.—A vegetable substance allied closely to cellulose. It occurs in irregular lumps, composed of granules which have a definite character, according to the variety of plant they were taken from. When dry its specific gravity is 1.53. Commercial Starch contains usually about 18 per cent. of water and, if kept in a damp place, will absorb 33 per cent. of water. It was much used formerly on solarized work. Torrefied Starch is obtained by roasting the common form, and is used in artificial leather compounds.

STIBNITE.—That ore of antimony known usually as black antimony. (See Kermes.)

SUBLIMED LEAD.—Used in the rubber manufacture, it acts both as a filler and chemically. Its peculiar velvety fineness makes it mix intimately with the rubber, and gives a very fine finish, showing no shiny crystals on the surface. The oxide of lead in the Sublimed Lead will also bind free sulphur in the rubber. The amorphous state of the Sublimed Lead makes the action of the lead oxide in this much more effective than the action of litharge, and the result is a very smooth lively jet black rubber.

SUGAR OF LEAD.—See Acetate of Lead.

SULPHATE OF LEAD.—A white powder of the specific gravity

of 6.2, insoluble in water, but readily soluble in caustic alkalies. It is not a very stable compound. In Cooley's formula for artificial leather, which has Gutta-percha for a base, it is used in connection with dextrine, magnesia, and cotton dust.

Sulphate of Lime.—Also called Gypsum. A common mineral occurring under various forms and names as alabaster, selenite, and gypsum earth. It is pure white in color and has a specific gravity of 2.33. Plaster of paris is a burnt form of gypsum. In the ordinary recovery of rubber by the acid process, whiting becomes gypsum. (See Anhydrite.)

Sulphate of Zinc.—Also called White Vitriol. It occurs in the form of a transparent crystal containing about 44 per cent. of water of crystalization, 87 per cent. of which is not given up short of a red heat. Its specific gravity is about 2.03.

Talc or French Talc is a mineral allied to mica. It is composed entirely of silica and magnesia, in the proportions of 67 to 73 of silica, 30 to 35 of magnesia, and 2 to 6 of water. Its colors are silvery white, greenish white, and green. Talc slate is more like steatite and is used for similar purposes. French Talc is used very largely in rubber factories in all lines of work for preventing surfaces from sticking together, during either manipulation or vulcanization. It is used also sometimes for dusting molds to prevent the gum from sticking to the metal and is used largely to bury white goods and keep them in shape during vulcanization. It is used sometimes in compounding, but any great amount of it produces a stony effect. It makes, however, an excellent semi-hard packing. It is used further in compounds for soft polishing, with India-rubber as a binding material.

Talite.—A white earthy material used in general rubber compounding. It is allied to diatomaceous earth, presumably, and has the same usage. Its analysis shows: Moisture 5.59, silica 83.9, sesqui-oxide of iron 1.2, alumina 2.8, oxide of manganese trace, potash trace, combined water and organic matter (by ignition) 6.47, loss and undetermined 0.04—total 100.

Tripoli.—See Rotten Stone and Infusorial Earth.

Wheat Flour is used in making matrices for rubber stamp work, and sometimes as a compounding material in India-rubber, though this is not to be advised, as the flour is apt to turn sour.

A large and important use for it has been in the dusting of black goods, such as rubber coats, so as to keep them from sticking together, should they accidentally touch during dry heat of vulcanization. Wheat Flour is preferable to almost anything else, for the reason that it washes off after vulcanization, without leaving any trace in color or stain. It is, of course, used on the goods known as "dull finished."

WHITING, or CHALK, as it is often called, is carbonate of lime. It is a white earthy material of the specific gravity of 2.7 to 2.9. It is made from English chalk, which is crushed, floated, and run through a filtering process, and dried in cakes, out of which, by a system of dry grinding and bolting, it is made in varying degrees of fineness. Where Whiting is kiln dried hastily, or under extreme heat, it is apt to become calcined, which gives it a hard, gritty feeling. Air dried whiting is considered the best. Whiting is in reality a purified form of carbonate of calcium, of a very soft or flocculent quality. The finest grades are known as "gilders'" and "extra gilders'." It is used more generally in rubber compounding than any other material, except sulphur. Used moderately, it increases the resiliency of rubber, but adds to the hardness. It does not, however, produce the stony effect that many ingredients give. It is also the basis of the molds used in rubber stamp making; paste being made of whiting, wheat flour, gule, and carbolic acid. Whiting is liable to absorb considerable quantities of water from the air. It is customary in many mills, therefore, to keep it in large bins that not only are covered but have steam pipes in the lower portions to drive out any moisture from the material.

WHITE LEAD.—This is a carbonate and is a heavy white powder. It is unstable in color, however, as sulphur compounds, especially in the gaseous forms, easily attack it and blacken it by reason of the formation of sulphide of lead. Its specific gravity is 6.46. Sometimes it is adulterated with lead sulphate, chalk, carbonate, or sulphate of baryta, or pipe clay. The simplest test for the purity of White Lead is to heat it in a thin glass vessel with some very dilute pure nitric acid; if pure it will dissolve completely. If chalk is present it also will pass into the solution, in which it may be detected by the addition of caustic potash, throwing it down as a white cloud. The best carbonate of lead is made by an old fashioned process, by placing metallic lead surrounded with spent tan bark in stacks, where it comes in contact with weak acetic acid. The heat of the bark volatilizes the acid and oxidizes the lead, while the acetic acid changes the oxide into acetate of lead, and this in turn is converted into carbonate by the carbonic acid given off by the heated body. This process of corrosion requires from six to eight weeks. There are many later and more rapid processes; for instance, take either litharge or acetate of lead, and expose them to a current of carbonic acid gas, etc. The original "triple compound" patented by Goodyear consisted of India-rubber, sulphur, and White Lead. A white lead known as sublimed lead is used very largely in the rubber manufacture. It is a fine white amorphous powder and imparts a decided toughness to rubber compounds. (See Sublimed Lead.)

UNUSUAL INGREDIENTS IN DRY MIXING.

It is not strictly accurate, perhaps, to say that it is unusual for fibers to be incorporated in rubber mixtures, for stocks made from unvulcanized rubber clippings have been used for years. Inner soles for rubber footwear and mats and molded articles have long been made of stocks of this kind, the fibers being cotton and wool, chiefly. Where wool was present there was oftentimes danger of blistering from the oil in the fiber, but this was easily gotten over by special compounding. In addition to the fibers already noted, silk, flax, jute and hemp—in fact, almost all of those in ordinary use—have been utilized, being added to the compounds to give toughness to them. The goods in which they are usually put are packings, artificial leathers, tire treads, and for wearing surfaces.

A fiber that has attracted considerable attention for this work, and one for which a number of patents have been granted, is cocoanut fiber, which is recommended for packings. Certain kinds of moss have also been used, as have sponge cuttings, peat, and wood pulp. This last named material has been used both in packings and in insulated wire compounds. It is also the basis of a curious artificial rubber that appeared several years ago, under the name of Maltha, but is not to be confused with the product that has become almost universally known by that uame.

Sawdust of all kinds has also been incorporated in rubber, and was formerly used in making sponge rubber, until better compounds were discovered. Those who use vegetable fibers prefer them unbleached rather than bleached, and very often treat them to remove resins that may be present. A few of the many other vegetable substances that have been used are sugar and sugar charcoal and seaweed. (See Algin.)

Animal substances are also valuable, as for instance, animal charcoal (which see), whalebone, which is called for in some of the Woodite patents, fur, tan-hair, leather fiber, Currier's skivings, which are used in artificial leather, the white of eggs, etc.

Under the head of earthy and metallic ingredients, almost anything can be used, although some metals have a bad effect on rubber, copper being the most notable of these. The unusual earthy matters are powdered fossil iron-stone, Wisconsin mineral, coke ashes, Stourbridge clay, powdered granite, salt, powdered lithographic stones, powdered oyster shells, powdered schist; and in metals, steel, and all other common metal borings, filings, and turnings. These latter have been incorporated in packings as a rule. One packing in particular, which has had a world-wide reputation, was heavily compounded with brass filings.

The deodorization of rubber, and the neutralization either of the smell of the rubber or its solvent, has brought out also a curious line of ingredients. Musk, for example, has been used to disguise the earthy odor of Gutta-percha. Alcoholic infusions of sage-tea, lavender, and verbena have been used in fine goods, while in powdered form, ginger root, birch, orris root, sassafras, marshmallow root, sandal wood, and other sweet smelling ingredients have been incorporated. The fossil of the mint has also been mingled with copperas, and placed in dry heaters, while a more expensive process was that pursued by Hill, who passed a current of hot air over perfumes and into the heaters. It must not be imagined that the ideas expressed in the foregoing are unworthy of the consideration of those who make ordinary cheap mechanical goods, for certain of these ingredients are used to-day in mechanical mixtures to overcome the odors of African rubbers. Essential oils and gums are also used for the same purposes, the descriptions of which will be found under their proper departments.

Medical science has also added its list of ingredients to rubber compounding, chiefly in the line of adhesive plasters, where ingredients like dry mustard, menthol, capsicum, belladonna, and a great variety of other medicaments are incorporated with the rubber.

CHAPTER VI.

I. SUBSTITUTES FOR INDIA-RUBBER AND GUTTA-PERCHA.

RUBBER SUBSTITUTES, as a rule, are made from oxidized oils. Those used most generally are made from linseed, rapeseed, cottonseed, mustard, peanut, or corn oils, acted on either by chloride of sulphur or by sulphur boiled with the oil at a high temperature. Substitutes have been known nearly fifty years, and have been made the subjects of many patents, but only within the last ten or fifteen years have they come into general use. French manufacturers have long exported these goods; they were really the first to produce them commercially. The fact that Europeans were unable at first to get the results with reclaimed rubber that were secured in the United States, led them to go further in their experiments with oxidized oils and to exploit their uses more thoroughly. The substitutes on the market to-day are, as a rule, white, brown, and black. They are slightly heavier than pure India-rubber, but their specific gravity is so near that of rubber that their presence cannot be detected in rubber compounds by specific gravity tests. Substitutes of this type are easily analyzed by the expert chemists, and the results of such analyses are of value to rubber manufacturers. The table on the next page, containing analyses of typical sorts of substitutes, is adapted from Dr. Rob. Henriques*.

It would be a mistake to suppose that rubber substitutes are of no value, for, as a matter of fact, they possess certain very distinct advantages not found in simple mineral adulterants nor possessed by any of the bituminous products now in use. Their value, of course, is where they cheapen stock without seriously injuring its durability or changing its texture. Among the wiser of the manufacturers, where substitutes are compounded with rubber they are used in small quantities, sometimes only 5 per cent. being added, and rarely is more than 25 per cent. to be found in a good compound.

Many substitutes, made from sulphurized drying oils, shorten the life of goods materially, by oxidizing the rubber. Manu-

---

**Journal of the Society of Chemical Industry*, 1894, page 47.

ANALYSES OF OIL SUBSTITUTES.

Oils Vulcanized with S_2Cl_2.	Sulphur.	Chlorine.	Water.	Residue on Ignition.	Fatty Acids.	Iodine Value.	Acetyl Value.	Fatty Acids.		
								Sulphur.	Chlorine.	Iodine Value.
Substitute from:	*Pr. Ct.*	*Pr. Ct.*	*Pr. Ct.*	*Pr. Ct.*	*Pr. Ct.*	*Pr. Ct.*	*Pr. Ct.*	*Pr. Ct.*	*Pr. Ct.*	*Pr. Ct.*
Linseed oil (raw oil)	9.34	8.84	3.02	....	79.6	56.3	21.0	9.88	Trace	160.3
Linseed oil (blown oil)	4.78	4.85	0.85	....	81.67	52.6	19.6	4.06	0.60	{ 141.2, 121.0
Rape oil (commercial oil)	8.28	7.62	....	....	86.89	32.5	31.0	8.34	Trace	101.5
Rape oil (blown oil)	6.59	5.95	....	....	87.95	26.9	....	6.54	Trace	102.8
Poppyseed oil (blown oil)	7.68	7.44	....	....	74.90	33.6	....	8.32	....	133.3
Cottonseed oil (blown oil)	6.23	5.36	....	....	....	30.3	51.3	6.44	Trace	91.5
Castor oil (with minimum S_2Cl_2)	4.82	6.70	....	....	85.35	35.2	....	5.32	Trace	{ 136.2, 147.4, 152.1
Castor oil (with maximum S_2Cl_2)	10.60	8.95	....	....	....	21.9	105.6	....	0.26	105.6
Commercial Products:										
White substitute No. 1	6.4	5.0	0.85	0.8	90.45	30.9	....	6.12	0.83	91.3
White substitute No. 2	6.17	5.86	1.0	5.51	73.58	31.0	....	6.45	0.43	91.2
White substitute No. 3	8.25	8.88	....	....	....	32.6	....	8.15	....	102.3
Brown substitute No. 1	15.48	0.7	....	....	....	42.0	....	14.14	....	129.0
Brown substitute No. 2	17.71	0.36	....	....	....	42.0	....	15.20	....	125.6

facturers have learned, however, to avoid those that have this fault, and are becoming more and more expert in the use of these goods, as they have become in the use of the cheaper grades of African rubber and reclaimed rubber.

The list following has been made quite comprehensive, not because all the substitutes described are deemed valuable, but rather to give a broad view of the subject. It will be noticed that many of these gums are far out of the line of sulphurized oil experiments. Resins, glues, asphalt, cellulose, seaweed, bastard rubbers, animal substances, etc., have all been called upon, and some of the treatments have been as original as the ingredients are unusual. To the end that the perfect substitute may be found, and with the fullest appreciation that anything which suggests new experiments has its value to the manufacturer, many that otherwise would be ignored are given here.

ADAMANTA.—The American name for a German substitute for India-rubber, made from linseed oil, sulphur, lime, and resin. It is a thick, black, gummy mass, with an odor similar to that of most of the sulphur oil substitutes, and showing a bright cleavage. It was at one time used largely in France and Germany, and introduced to some extent into the United States. Its chief use was in cheap mechanical rubber goods, and for insulation.

ALGIN GUM.—A gluey, leathery substance, manufactured from seaweed. It is insoluble in cold water, alcohol, ether, and glycerine, and combines readily with alkaline and metallic bases to form substances, many of which are soluble. Algin can be used for waterproofing compounds, as it combines easily with rubber, shellac, and other gums. With many metallic bases it forms insoluble compounds as tough as horn or as pliable as Gutta- percha. It is an English product.

A. R. D. GUM.—So called because it is used as an anti-dry-rot compound. It is manufactured of 112 parts glue, 56 parts resin, 10 parts boiled linseed, and 35 parts water. In some cases it has also been mixed with India-rubber in general compounding. Patented by J. F. Ebner, London, England.

ARTIFICIAL ELATERITE.—Made from liquid bitumen by incorporating with it vegetable oils, such as cottonseed oil, palm oil, rapeseed oil, etc. The product is treated with aid of heat and pressure, with chloride of sulphur, saltpeter, and sulphur, which produces an oxidization of the fatty substances. The result is an elastic rubber-like or leathery mass, which is soft, spongy, and gluey. This gum is said to be far more elastic than the best samples of mineral rubber, and is useful for waterproofing and insulation. Patented by W. Brierly, in England.

ARTIFICIAL GUTTA-PERCHA.—A French compound made of 50 parts copal, 15 parts sulphur, 30 parts turpentine, and 60 parts petroleum. While mixing the heat reaches 100° C.; it is then cooled to 35° C. Then there is added a solution of 3 parts caseine, in weak amonia, and a little methylene, and reheated to 120° C. It is then boiled with a 15 or 20 per cent. solution of tannin, and 15 parts ammonia. After several hours' boiling it is washed and cooled.

BLACK GERMAN SUBSTITUTE.—Made of boiled linseed oil and sulphur, together with resinate of lime. This gum is similar to Adamanta, and has been practically driven out of the market by lighter substitutes.

BLANDITE.—An artificial India-rubber invented by Dr. A. L. Blandy, of London. It is fairly elastic, stretching to about twice its length, and returning readily. It is very pliable and does not show signs of cracking when bent. It is vulcanized like ordinary rubber, and can be molded into any form desired. Coated on cloth, it strongly resembles leather. It is waterproof, and is used for gas tubing, mats, etc. In its crude form, it is a liquid mass resembling molasses. Dr. Blandy's patent describes the compound as made preferably of linseed oil which has been reduced by oxidation; then 10 per cent. of bisulphide of carbon, to which has been added 10 per cent. of chloride of sulphur, is mingled with the oil, and brought by gentle heating to the desired consistency. Trinidad asphalt, cleansed and reduced to powder, is combined under the heat in the proportion of 3 parts to 1 of oil. Care must be taken to avoid fire in heating. These proportions are gradually brought, by heat and stirring, to a liquid or thin state, and when in this condition it must be poured upon a wet, cold surface, and thus cast into sheets, convenient for subsequent mixings.

CARROL GUM.—A well-known sulphur oil substitute used in

the United States. In smell it has all of the characteristics of the sulphurized oil products. It is produced usually in granular form, and is very black.

CHRISTIA GUM.—An English substitute for Gutta-percha or India-rubber, used as a surgical dressing. It is said to be composed of hemp fibers, so treated as to be impervious to both alcohol and water. Dieterich analyzed a sample of the product, and said that the fibers were sulphite wood pulp, and that the coating was made from chrome gelatine treated with glycerine, or the well known compound of glue, glycerine, and bichromate of potassium.

CORKALINE is made of glue, glycerine, ground cork, and chromic and tannic acids. It is of English derivation and is used as a substitute in mat work.

CORN OIL SUBSTITUTE.—A sulphurized oil substitute similar to that made from oxidized linseed or rapeseed oils, manufactured from corn or maize oil. It is the cheapest oil substitute that has yet been put on the market. It is made in two colors, brown and straw color, and is used in large quantities in mechanical goods, and in proofing. A good example of this type of substitute is that known on the market as "Kommoid."

DANKWERTH'S RUSSIAN SUBSTITUTE.—This is said to be a perfect substitute for both Gutta-percha and India-rubber, and is used for covering telegraph cables. High temperatures do not affect it. It is made of 1 part by weight of the mixture of equal parts of wood tar, oil, and coal tar oil, with 2 parts of hemp oil heated until the mass is of the right consistency. Then 1-3 part by weight of boiled linseed oil is added. To this is added a little ozocerite and some spermaceti. It is then heated again, and finally a little sulphur is added.

ELASTEINE.—An elastic substance produced through the treatment of certain resins. Solid and semi-solid copal resins are treated with oleic acid (found in stearine works), which entirely dissolves them. The product of the solution is soluble in spirits of turpentine and in oil. This solution of gums in oleic acid gives an opportunity to produce materials that have sometimes the elasticity and the consistency of India-rubber. The inventor advises their use in insulating wire and in various kinds of proofing. It is of French origin, and patented by M. Louis Riviere.

Elastic Glue.—A mixture of dry glue and glycerine in equal parts, by weight. As little water should be used as possible in its manufacture. It is used for elastic figures, galvano-plastic molds, etc. It is not waterproof, nor will it stand a high degree of heat.

Euphorbia Rubber.—J. G. Boles reduced euphorbia gum to a fine powder and, after drying carefully at a low temperature, put it in solution and finally hardened it by mixing it with earthy matters and shellac. The same gum before that he mixed with a preparation of rubber and cured it, forming a kind of vulcanite.

French Gutta-percha.—This gum is made by boiling the outer bark of the birch tree in water. The result is a fluid, which is very black, and which becomes compact and solid on cooling. It has been claimed that it possesses all of the good properties of Gutta-percha, and that in addition it does not oxidize when exposed to the air. Its application for industrial purposes has been patented.

Fenton's Artificial India-rubber.—Manufactured from linseed or similar oils, mixed with tar, pitch, or other forms of pyroligneous acid, the mixture being placed in a bath of diluted nitric acid, and allowed to remain for maceration until, by the action of the bath upon the compound, the whole is coagulated into a tough, elastic magna. The black "Fenton" contains as a coloring matter a small quantity of plumbago or black carbonate of iron. The gum is patented by Ferrar Fenton, London, England. In his specification he modifies it by taking the artificial gum described, and placing it in a bath composed of a solution of sugar of lead, oxide of zinc, saltpeter, or some other form of nitrate, and, if high flexibility is desired, adds 5 to 10 per cent. to copal gum and nitric acid diluted with water. These solutions are used one at a time, the proportion being 5 per cent. of sugar of lead, or, for greater hardness, 5 to 7½ per cent. of saltpeter to the weight of the magma. Before vulcanizing, the substances are washed in an alkaline solution to remove acid. Fenton rubber is said to have been subjected to 320° F. for fifteen minutes, the only result being to increase its elasticity.

Grape Rubber.—A high grade of artificial rubber, produced from the skins and seeds of grapes from which wine has been extracted by pressure. Small samples manufactured in the laboratory are said to be almost identical with pure rubber. It has been impossible so far to make the material on a large scale economically and, therefore, none of the gum is on the market.

Gum Fibrine is made of paper rags, treated with liquid carbonic acid, mixed with resin and gum benzoin and castor oil, dissolved in methylated alcohol. It is an English compound.

Guttaline.—A substitute for India-rubber and Gutta-percha, manufactured as follows: To Manila gum tempered with benzine is added 5 per cent. of Auvergne bitumen, also mixed with benzine. Then add 5 per cent. of resin oil, and allow 48 to 86 hours to pass between treatments. The product obtained is similar to India-rubber. If it is too fluid, the addition of 4 per cent. of sulphur dissolved in bisulphide of carbon will act as a remedy.

Insulite.—A preparation made of wood or vegetable fiber, finely ground and dessicated, and saturated with a mixture consisting of melted asphalt, incorporated with substances of the resin type, with or without substances of the paraffine or anthracine types. The products resulting are used as substitutes for India-rubber, particularly in insulation. Patented by Alfred H. Huth, London.

Kelgum.—A linseed oil preparation manufactured in the following way: First, boiling linseed oil in a nitric acid bath until it reaches a gum-like condition; second, subjecting the gum to a bath for the removal of the acid; third, cutting the gum in a solvent bath; fourth, disintegrating the gum with the solvent; fifth, grinding the disintegrated mass; sixth, boiling the material; seventh, subjecting the same to another boiling, and adding a drier. Used in proofing compounds. Invented by Henry Kellog, United States.

Kerite.—A compound of vegetable oils, coal tar, bitumen, and sulphur, to which is added sometimes a little camphor and various waxes. Occasionally sulphide of antimony is used in place of sulphur. Vegetable astringents, such as tannin, the extract of oak bark, etc., are also used in small quantities to impart toughness. Kerite is the invention of Austin G. Day, and has been used largely for the manufacture of a covering for insulated wire.

Kommoid.—See Corn Oil Substitute.

Linoxin.—An insoluble oxy-compound produced by the oxidation of certain drying oils boiled in acetone or acetic acid, from which is produced an elastic mass similar to India-rubber. Of French origin.

LUGO RUBBER.—An artificial oxidized oil substitute that originated with a German chemist, Dr. Lugo, who introduced it into the United States, where it once had a large sale. It was black, of about the same specific gravity as India-rubber, and made, in connection with rubber, excellent mold work. It is not now on the market.

MAPONITE.—A substitute for India-rubber and Gutta-percha, claimed to be capable of use in the manufacture of golf balls, tobacco pouches, etc. It is said to be vulcanizable at 260° F. An English patent has been applied for, the inventor being F. E. MacMahon.

NIGRUM ELASTICUM.—A sulphurized oil substance apparently made from linseed oil. Very dark colored and quite hard. Of English origin.

NOVELTY RUBBER.—An English substitute invented by David Lang. It is made red and drab in color. It comes in small slabs about 18 inches square and 2 inches thick, weighing about 7 pounds. It is said to be easily mixed with ordinary rubber, vulcanized in the usual way, the price being about the same as for reclaimed rubber.

OXOLIN.—An English invention patented by Charles J. Grist, an electrical engineer, and identical with "Perchoid" in the United States. This gum is used for waterproof sheeting, printers' blankets, packings, etc. It is made of a solution of partially oxidized oil by adding litharge and heating to over 400° F. Jute, or other fibers, is then dipped in the oil, the surplus oil is removed in a hydro-extractor, and the oil remaining on the fibers is oxidized by a current of air. These operations are repeated twice. The material is then ground with sulphur and coloring matters, and treated like India-rubber.

PARKESINE.—Made from a compound of linseed oil and pyroxyline, and used in the manufacture of small articles that are sometimes made of hard rubber. A Parkesine compound for molding, proofing, etc., is as follows: To 500 pounds water add 50 pounds sulphuric acid, and steep in it as much cotton, or rags, or jute, or linen as the liquor will moisten, for 3 or 4 hours. Take out, drain, and expose the mass to steam heat of about 280° F., for an hour, if cotton or jute fiber has been used, and 3 hours if flax. Neutralize the acid pulp with a bath of water and soda, using 4 pounds of carbonate of soda to every 200 pounds of rags. Wash and press, pass through a coarse sieve of 12 meshes per inch, and dry. Grind the granulated material and sift it through a sieve of 120 meshes to the inch. The resulting powder may be mixed, in all proportions up to equal parts, with fresh rubber. Compounding 25 to 50 parts dry Parkesine, with 50 parts alcoholic solvent. A proofing compound is: 1 pound paraffine, linseed oil, or other drying oil; 4 to 8 ounces Parkesine.

PERCHOID.—See Oxolin.

PEROXIDE SUBSTITUTES.—Peroxide of lead having been recommended as a better drier than other oxides used in connection with all compounds, the following formulas are given: 25 parts of walnut oil, 62 parts linseed oil, 5.5 parts peroxide of lead, 7.5 parts sulphur. One of greater toughness is composed of 25 parts walnut oil, 56 parts linseed oil, 5 parts peroxide of lead, 6 parts sulphur, 6 parts gum juniper. [Prof. W. Lascelles-Scott.]

PICKEUM SUBSTITUTE.—This is made by the following treatment of Pickeum gum:

A.

Boiled linseed oil	160 pounds.
Vaseline	20 pounds.
Bastard gum (or Pickeum gum) from Central America, cut fine	40 pounds.

Stir and heat to 250° to 300° F., until the gum is dissolved. Then cool to 100° F., and strain.

B.

Mixture of	Solution as above	5 gallons.
	Protochloride of sulphur	9 pounds.
	Bisulphide of carbon	9 pounds.

After the chemical action takes place, the mass is granulated and the grains are washed and stored for use, or the material may be masticated in a rubber mill and run into sheets for use.

PURCELLITE.—The invention of Dr. C. Purcell Taylor, of England. An insulating substance somewhat similar to Gutta-percha, but costing much less. It is said to be very tough and elastic, may be made of any color, and is either flexible or rigid. The specific gravity of the material is 1.2. It can be molded or vulcanized like India-rubber. Its insulation resistance is equal to that of Gutta-percha. It is unaffected by atmosphere, by alkaline or acid liquids, freezing mixtures and the like.

RESINOLINES.—Substances so called by Eugene Cadoret, of Paris, who obtains them by saponifying various oils by the use of a metallic carbonate, using by preference carbonate of lead, then decomposing by nitric acid, decanting, and saturating with an alkali. The soap thus formed is treated with acid to form a resinoid body, purified by dissolving in alcohol, and evaporating the solution. Resinolines thus formed are very similar to natural resins. They are either semi-fluid, pasty, or solid. When solid, they are remarkable for their flexibility.

ROSALINE.—A vegetable product said to contain about the same chemical elements as India-rubber, and of about the same specific gravity. Manufactured in United States, France, and England. A strong point is made by the manufacturers that after vulcanization no chemist is able to detect that there is anything but pure rubber in a mixture containing 25 per cent. of Rosaline and 75 per cent. of India-rubber. In vulcanizing, it requires about one-third more time to bring about the usual result.

RUBERINE.—An American rubber-like solution used as an insulating paint, and also as a proofing mixture, and partaking of many of the qualities of ruberoid. It is also manufactured in Germany.

RUBEROID.—An American substitute for India-rubber that has the physical appearance of a high grade of black oil substitute. In use, however, it differs from many of them, for the reason that it has been found useful in vulcanite compounds, while at the same time it may be used in ordinary soft rubber work.

RUBBERITE.—An artificial rubber of the same specific gravity as fine Para. In color, elasticity, capability for vulcanization, and durability, it is said to resemble the higher grades of rubber. It is the invention of H. C. B. Graves, London, and is made up as follows:

Trinidad asphalt.............................47 to 80 per cent.
Oxidized oil.................................20 to 30 per cent.

Vaseline	5 per cent.
Sulphur......................................	15 per cent.
Chloride of sulphur	3 per cent.

RUBBERAID.—An amber colored substitute manufactured from cottonseed oil by a secret process, which removes what the inventor calls the grease, leaving an elastic semi-solid which has been used quite largely in compounding.

RUSSIAN SUBSTITUTE.—Manufactured from the skins of rabbits and other small animals, or the waste therefrom, digested in crude glycerine, and a little water. The formula is 3 parts by weight of the cleansed substance melted in water, with 3 parts by weight of crude glycerine, to which is added ¼ part by weight of a concentrated solution of potassium chromate. The resultant mass is flexible. To make it harder, a little less glycerine and more chromate of potash are required. To withstand acids, 30 per cent. of gum lac dissolved in alcohol is added. For waterproofing fabrics, ¼ part by weight of oxgall is added, with enough salt water to give it the consistency of oil.

SOAP SUBSTITUTES.—These have been exploited and explained more thoroughly by Prof. W. Lascelles-Scott than by anybody else. The typical formulas that he gives are as follows: 28 parts of aluminum soap, 60 parts of linseed oil, 8 parts of acid free sulphur, 4 parts of oil of turpentine. Another, to use in connection with reclaimed rubber, is 15 parts of aluminum soap, 25 parts of devulcanized rubber, 60 parts fresh rubber, benzine *quantum sufficit*. Another still, in which a low grade pseudo gutta is used, is 15 parts aluminum soap, 25 parts Almadina gum, 5 parts raw rubber, 6 parts sulphur, and 4 parts oleum succini.

TEXTILOID.—A mixture of a resinoline [as described by Cadoret under that heading] with natural resins, cellulose, nitric cellulose, or organic substances of animal origin. The resultant material may be transparent, white, or colored. It is practically uninflammable, has no smell, is very elastic, and, if submitted to heat, softens, and can be easily drawn out into fine threads. It can be used for waterproofing and in various other ways is a good substitute for India-rubber. It is flexible and elastic. Textiloid is made of 4 parts resinoline, 2 parts nitric cellulose, and 1 part camphor dissolved in alcohol at 90° F. The result thus formed may be made in colors by the addition of metallic oxides.

TONG OIL SUBSTITUTES.—Manufactured from the Chinese oil known as tong oil, or wood oil. The oil is heated without any foreign matter being added to it, at a temperature of 250° C., when it becomes solidified. It is then pulverized, and impregnated with petroleum, which swells it, and renders it more easily worked. Patented by Dr. Charles Repin, Paris.

TURPENTINE RUBBER.—Manufactured by passing spirits of turpentine through a heated tube so as to vaporize it, and mixing the vapor with hydrochloric or other acid, so as condense and solidify all of the vapor. Patented by A. F. St. George, England.

TREMENOL.—A German invention that has reference to the production of sulphonic acids, sulphones, oils, resin oils, mineral waxes, etc. Results from a treatment of mineral matter with fuming sulphuric acids at ordinary temperatures, or with concentrated sulphuric acid at 120° C. The invention further calls for the treating similarly of the bodies obtained from the oil in their precipitation by means of sulphuric acid. The products are then washed in brine and water. The inventors precipitate glue and gelatine from a slightly acid solution, as elastic rubber-like substances that can be drawn into threads with perfect ease.

VOLTIT.—The base of this is glue or gelatine prepared from scraps of kid skins, which are treated until they reach a gelatinous mass, which is filtered and mixed with oleic acid, such as is used in candle factories, the proportion being 80 parts of oleic acid to 20 parts of the gelatine. The mixture is boiled for ½ hour, and then 11 parts of caustic potash solution (in 50 parts of water) is added. The boiling is then continued for an hour, and a special mass is formed to which is added resin oil, oxidized linseed oil, and paraffine. The whole mixture is then boiled 4 to 5 hours. Also spelled Voltite. It is of French origin.

VOLENITE.—A substitute for India-rubber and Gutta-percha invented by Frederick Lamplough, United States. The compound is said to be a mixture of resins, or resin oil conveyed into a mass of fibrous material by a suitable non-oxidizable oil. This latter oil is used simply as a vehicle to carry the resin to its place, the process being completed by the distillation of the non-oxidizable oil, and the oxidizing of the rest of the mass. The oil used is preferably a fish oil, which is refined carefully before use. After saturation and treatment the vegetable fiber is changed into a homogeneous mass which has many of the characteristics of vulcanite. A formula that is said to have worked well is 10 parts by weight of fiber, 5 parts resin, 4 parts resin oil, 2 parts fish oil, treated at a temperature of 130° C., for 4½ hours.

WATERPROOF GLUE.—A substitute for canvas proofing made as follows: Dissolve 16 ounces of glue in 3 pints of skim milk, and to increase its strength add a little powdered lime.

WINTHROP GUM.—Another name for Rubberaid.

II. SUBSTITUTES FOR HARD RUBBER AND GUTTA-PERCHA.

HARD RUBBER in its best estate is so valuable and perfect a product that it would always have the preference were it not for its unavoidable high cost. Because of this cost there are many substitutes for it that counterfeit it in texture, color, and quality, but are never quite its equal in all these points of excellence. These substitutes are made of cellulose, gums, and animal, vegetable, and earthy matters, having a variety of distinctive names and varied uses. To the popular mind, if they look like ebonite, they are hard rubber. In the same way, Gutta-percha is often confounded with hard rubber, which it resembles under many conditions. The following list covers not only certain widely-known compounds of hard rubber and Gutta-percha, but a number of substitutes for them now put to many uses, the chief of which, perhaps, is insulation:

ALEXITE.—An American insulating material which can be molded in any shape, is waterproof, fireproof, and acid proof, and can be produced in any color. In texture and general appearance it resembles vulcanite.

AMBROIN.—A German substitute for hard rubber, consisting of fiber, silica, and resin compressed to a mass. Its color varies from light brown to green or black. Nitric and acetic acids do not effect it, and even aqua regia does not injure it. Under a moderate heat it softens slightly and can be worked, like vulcanite, in a mold. It also takes a bright finish from the buffing wheel.

ARMALAC.—See Insulac.

ARTIFICIAL WHALEBONE.—A well-known product made as follows: India-rubber 20 parts, sulphur 5 parts, shellac 4 parts, magnesia 4 parts, and gold brimstone 5 parts. Vulcanized somewhat the same as hard rubber.

BALENITE, as the name signifies, is intended as a substitute for whalebone. It is quite elastic; in other words, it is neither hard nor soft, but may be characterized as semi-hard. A well-known compound for this is India-rubber 100 parts, shellac 20 parts, burned magnesia, 20 parts, sulphur 25 parts, and orpiment 20 parts. (Hoffer.)

BITITE.—An English insulating material which is said to be bitumen refined to absolute purity and vulcanized. It is used on cables, in underground work, for low pressure resistance, and in rare instances for high pressure.

BROOKSITE.—A compound of resin and heavy resin oils for insulating purposes.

CAOUTCHOUC ALUTA.—A composition used as a substitute for hard rubber, made of leather scraps boiled in water, with a sufficient quantity of oxalic acid to dissolve them, and a portion of glue. To this are added resin, pitch, beeswax, and copal gum, dissolved in oil. India-rubber boiled in linseed oil is then added and a powder formed of plaster of paris, and a coloring matter is stirred into the composition to thicken and stiffen it.

CHATTERTON'S COMPOUND.—A widely-known compound sold the world over for connections for joint sheets and for uniting Gutta-percha parts, and also used for cementing Gutta-percha to wood. It softens readily at 100° F., and becomes firm again when cold. Its specific gravity is about 1.02. The best compound is 1 part by weight of Stockholm tar, 1 part resin, and 3 parts cleansed Gutta-percha, melted and mixed.

CORALITE.—A name for vulcanite which is colored to imitate coral.

CORNITE.—A specially hard vulcanite or hard rubber, so named from the Latin *cornu* (a horn).

DIATITE.—A combination of diatomaceous earth, and shellac, made under very heavy pressure. It may be made of any color, and is used as a substitute for hard rubber.

ELECTROSE.—A substitute for hard rubber for which the following advantages are claimed: It will not tarnish metal, as no sulphur is used in its vulcanization; it is cheaper than hard rubber; it possesses high insulation properties; it can be melted readily into any shape, or made of any color; it does not fade; it possesses great strength, and takes a high polish; changes of temperature do not affect it; and it withstands the weaker acids and alkalies.

ESBENITE.—Made of pure cellulose, chemically incorporated with mica in the form of fine powder, with the addition of magnesia and silicate, thus forming strong and close grained artificial mica. It is flexible, and can be molded into any shape. Esbenite is waterproof, does not burn readily, and is thoroughly airproof. Manufactured in England.

FIBRONE.—A substitute for hard rubber which is a good non-conductor, waterproof, and can be handled in a lathe like vulcanite. It is said to be durable, does not contract or expand, and is made in all colors. It is used for thumbscrews, pushbuttons, etc. Plasticon is similar to Fibrone, but heavier and of a more stony nature, and probably made of the same material.

HYALINE.—Made of a mixture of equal parts of gun cotton and a variety of resins. The gun cotton is dissolved in ether and the resins in solution are added, the result being a thick, gelatinous mass. When allowed to dry, this mass soon hardens and forms a horny, incombustible material. Invented by Frederick Eckstein, Vienna.

INSULLAC.—A spirit copal resin varnish, with the acids of the resins neutralized as much as possible, to prevent the resin acids from attacking the copper wire. It is a transparent elastic material, and is superior to shellac. Armalac is made of black paraffine wax, in solution in petroleum. It remains permanently plastic under heat, although it dries quickly and thoroughly. Manufactured in the United States.

INSOLACIT.—An insulating material produced either as a liquid, semi-liquid, or solid. It is not inflammable or affected by the most corrosive acids, alkalies, saline substances, etc. It is a German product and the compound remains a secret.

ISOLATINE.—An American insulating material prepared especially for high resistances. It is said to be very flexible, not to be affected by cold or heat unless the latter is artificial, and to be very durable. It is also said to protect metal.

KIEL COMPOUNDS.—One of these well-known compositions consists of India-rubber, sulphur, pumice stone, oil, and beeswax. The resultant compound makes a hard rubber, said to possess a superior elasticity and toughness, and capable of being vulcanized in sheets at least 2½ inches thick. This compound is not affected by the most intense cold, and will stand a higher temperature than ordinary rubber. It also burns with difficulty. Its ingredients are said to mix faster and more uniformly than those of other compounds. It resists acids, and other corrosive substances, is a perfect insulating material, and is cheap. Another Kiel compound is made of India-rubber, sulphur, and mineral oil. The resultant compound is more flexible than ordinary hard rubber, and when warm is more plastic than such compounds. It is also less brittle and cheaper, and can be turned in a lathe with greater facility and less injury to the tools.

KERATITE.—Another name for hard rubber, derived from the Greek word meaning horn.

KERATOL.—An American waterproof preparation, not of the nature of rubber, but probably one of the cellulose substitutes. It is a colorless transparent substance, and when applied to fabrics renders them waterproof and prevents crocking and fading. It also strengthens the fabric, and allows stains to be washed off. An artificial leather is also made of Keratol. The name is adapted from the Greek word *keros,* meaning hornlike. Invented by Parker R. Bradley, United States.

LAMINA FIBER.—An American invention, used chiefly for electrical purposes. It is of various colors, heavier than vulcanized rubber, and swells to nearly double its weight when placed in water. It is probably a cellulose compound containing no rubber.

LACTITIS.—An artificial ivory made from milk, the process being coagulation, straining, and rejection of the whey. Ten pounds of the curd are then taken and mixed with the solution of 3 pounds of borax in 3 quarts of water. The mixture is then placed in a vessel over slow fire and left until it separates into two parts, one as thin as water, the other resembling melted gelatine. The watery part is drawn off, and to the residue is added a solution of 1 pound of mineral salt in 3 pints of water. (Sugar of lead answers very well as the mineral salt.) This brings about another separation of the mass, into a liquid and a mushy solid. The liquid is strained or filtered off, and at this point coloring matter may be added. The solid is now subjected to heavy pressure in molds of any shape, and afterwards dried under great heat. The resulting product may be used in the manufacture of billard balls, knife handles, or anything for which ebonite or celluloid is adapted.

LEATHEROID.—A mixture of American origin, made in black, red and gray, and similar to vulcanized fiber. It is insoluble in ordinary solvents, uninjured by alcohol, ether, ammonia, turpentine, naphtha, or other oils, is very tough, is a good insulator, and is of low cost.

MARLOID.—An insulating material said to be made from the hides of certain animals, treated by a chemical process, making it so hard that it can be handled in every way the same as ebonite. It may be transparent or opaque, and is capable of receiving a very high polish. It is said to give an insulation of 2,000 megohms, is uninflammable, and is of English origin.

MICANITE.—Mica cemented together under pressure with an India-rubber compound. Manufactured in America.

NIGRITE.—An insulating compound consisting of a mixture of India-rubber and ozocerite.

PEGAMOID.—This, although covered by several patents, is said also to involve certain secret processes. In a general way, however, the substance is prepared by treating a fine grade of cellulose with a mixture of sulphuric or nitric acid to form nitrocellulose or gun cotton, which is then dissolved in a suitable alcohol. The Pegamoid patents call for the addition of glycerine, sweet or olive oil, and various coloring matters.

PLASTICON.—See Fibrone.

PLASTITE.—A vulcanite which is made extra hard and is not possessed of any special amount of elasticity. The stock recipe for this is: India-rubber 100 parts, sulphur 25 parts, magnesia 50 parts, orpiment 50 parts, coal tar asphaltum 60 parts. It is very hard and solid, and takes a high degree of smoothness and polish. (Hoffer.)

POTATO CELLULOID.—An Austrian invention relating to an artificial solid produced from potatoes boiled 36 hours in a fluid containing 8 parts of sulphuric acid and 100 parts of water, and then dried. Pipe bowls made from it for the French market are said to be hardly distinguishable from real meerschaum. Billiard balls are also said to be made from it.

PRESSPAHM.—An English insulating material made from wood fiber so treated that it can be run through rolls into sheets of varying thicknesses. It is said to be capable of withstanding high temperatures, and is used not only in connection with electrical machinery, but also for bookbinding and for putting a finish on cloth.

SOREL'S COMPOUND.—A so-called substitute for Gutta-percha consisting of 2 parts resin, 2 parts asphaltum, 8 parts resin oil, 6 parts slaked lime, 3 parts water, 10 parts potter's clay, and 12 parts Gutta-percha. Five per cent. of stearic acid is sometimes added.

STABILIT.—A German invention, the compound for which is a secret, designed to be half way between hard rubber and vulcanized fiber. It is not affected by corroding substances, and does not absorb moisture. It withstands boiling water where hard rubber and vulcanized fiber do not, and is not attacked by muriatic acid or sulphuric acid.

VEGETALINE.—Cellulose treated with sulphuric acid, dried and ground, and then treated with resinate of soda.

VISCOSE.—An English cellulose product that promises much a substitute for vulcanite. It may be of any color or any degree of hardness. It has been used in connection with rubber experimentally with excellent results. As a friction for belting it is said to be excellent, whether or not the belt has the regulation rubber cover.

VISCOID.—A compound of viscose, formed by mixing with it hot bituminous matter such as tar, pitch dissolved in coal tar, or the like. The resultant mixture, when solidified, constitutes a material of a high insulating character, and is produced at low cost. The bituminous and cellulose matter may be mixed in equal proportions, although there is a wide range of compounds that may be made through the use of various proportions of the substances.

VITRITE.—A jet black, perfectly hard material, having a smooth polished appearance similar to ebonite. It is not affected by dampness or acids. It is a good insulator, is of low cost, and easily worked.

VULCABESTON.—This is a composition of asbestos and India-rubber, forming a product which is a non-conductor of electricity and stands the severest tests, resisting heat wonderfully. Invented by R. N. Pratt, United States.

VULCANIZED FIBER.—This material, which is very largely used, is made of cotton paper pulp, chemically dissolved, and solidified under enormous pressure. It is unattacked by ordinary solvents such as alcohol, turpentine, ammonia, etc. It appears on the market in two forms—hard and flexible. The hard fiber resembles horn and is exceedingly tough and strong, while the flexible fiber has the appearance of a very close grained leather. It is an insulator in dry places, but, as it will absorb moisture, it is useless in places requiring waterproof qualities. It is made in three colors—black, red, and gray. Vulcanized fiber is unaffected by oils or fats, and will stand action of hot grease. Low grades have been found adulterated with chloride of zinc and calcium, to the extent of nearly 50 per cent. of its weight.

WILLOUGHBY SMITH'S GUTTA-PERCHA.—Gutta-percha refined by a special process invented by Willoughby Smith. Valued in England as giving an increased speed over electrical conductors insulated with it.

WRAY'S COMPOUND.—A composition of India-rubber, silicia, powdered alum, and Gutta-percha. Used in climates too hot for Gutta-percha by itself. It is easily attacked by seawater.

III. MISCELLANEOUS SUBSTITUTES AND COMPOUNDS.

"APO ELASTIKON HYPHASMA."—An English formula for this is: Take caoutchouc and grind it with a portion of residue from cottonseed oil. Work in as much vegetable fiber as will convert it into a strong felt, adding as much farinaceous matter as will fit it for the finishing roller. The outside husk from rice, finely ground, is preferable. Stearine pitch may be added to give a greater stiffness; also chalk and steatite may be used to harden it.

ASBESTONIT.—An asbestos product manufactured in England under a secret process, for use as steam or hot water packing.

ASTRICTUM.—A compound to be used in damp places, consisting of pulped cotton 15 pounds, pitch 25 pounds, asphalt 20 pounds, ground granite rock 20 pounds, bitumen 5 pounds, resin 10 pounds, coal tar 12 pounds, and mastic 5 pounds.

CAOUTCHITE.—Vulcanized rubber exposed to heat (250° F.) for several days and devulcanized and recovered by this means alone.

CORK LEATHER.—A French invention composed of thin sheets of cork, covered on both sides with an extremely thin India-rubber skin, and of a textile fabric outside. It is very light, is a good insulator against heat, and is waterproof.

DERMATINE.—A well known substitute for India-rubber and leather, made of an artificial Gutta-percha called "gum percha," 7 pounds; powdered waste rubber, 7 pounds; India-rubber, 14 pounds; sulphide of antimony, 6 pounds; peroxide of iron, 2 pounds; flour of sulphur, 2 pounds 8 ounces; alum, 4 pounds 8 ounces; asbestos, powder, 8 pounds; sulphuret of zinc, 3 pounds; carbonate of magnesia, 7 pounds. A little change in this compound adapts it for machine belts. A variety of colors is gained by mixing in various pigments in place of sulphuret of antimony or peroxide of iron. The invention is patented by Maximilian Zingler, of London. It is claimed that Dermatine will stand more wear than either leather or rubber, that it is absolutely unaffected by heat, cold, dryness, or moisture; and that it will stand perfectly the action of grease, oils, or acids. Adaptations of the formula given above permit it to be manufactured in molded forms. It is used for valves, packing, etc., and also for covering insulated wire.

DURATE.—An artificial rubber compound said to be similar to Dermatine.

FIBRINE-CHRISTIA GUM is manufactured just as Christia gum is, except that silk fibers are used in the place of hemp.

FROST RUBBER.—Another name for what is practically sponge rubber made from any ordinary unvulcanized rubber compound by the addition of a little alum or carbonate of ammonia.

HEVEENOID.—This is claimed to be more insoluble, durable, and pliable than almost any other rubber composition. Soft Heveenoid consists of India-rubber 32 parts, camphor 32 parts, lime 1 part, and sulphur 8 parts. Hard Heveenoid is made of India-rubber 6 parts, camphor 4 parts, glycerine 1 part, and sulphur 16 parts. Heveenoid is the invention of Henry Gerner, of New York, and is patented in the United States and Europe. Kauri gum is also used in certain Heveenoid compositions. One special advantage claimed as to the use of camphor is that the chemical compound termed sulphide of camphor is produced, and therefore the rubber does not bloom.

HEVEENITE.—Another name for Heveenoid.

INDIA-RUBBER LEATHER.—A compound produced by Nelson Goodyear in which fibrous substances were mixed with India-rubber to form a body the surface of which resembles leather.

KAMPTULICON.—An India-rubber compound for floor coverings. The simplest English formula is a vegetable fibrous material ground into a coarse powder, mixed with India-rubber, and treated with a cheap solvent, such as coal tar or naphtha. Coloring matters are added, if desired. Another Kamptulicon compound is: Gutta-percha, cheap grade, 6 pounds; reclaimed rubber, 12 pounds; residuum from distilling palm oil, 6 pounds; ground cork, 4 pounds; ground chalk, 2 pounds; sulphur, 6 pounds; hair, 1 pound; oxide of zinc, 1 pound.

KIRRAGE COMPOUND.—A well-known English patented compound, which takes its name from the inventor. It comes in two forms. The first, to be used not over 200° F., is composed of India-rubber 12 pounds, Gutta-percha 4 pounds, Stockholm tar 25 pounds, chalk 60 pounds, hemp 4 pounds, and sulphur 10 pounds. The same inventor also recommends the following, to withstand a great heat and pressure: India-rubber 20 pounds, tar 25 pounds, coke, finely powdered, 25 pounds; Stourbridge clay 25 pounds, sulphur 10 pounds, fine emery 25 pounds, and steel filings 5 pounds.

LEATHERINE is a compound that closely approaches Dermatine, and in fact is a part of the first patent on that product. It is intended as a substitute for leather cloth and is made as follows: India-rubber 28 pounds, substitute 10 pounds, sulphuret of antimony 13 pounds, peroxide of iron 4 pounds, sulphur 3 pounds, sulphuret of zinc 10 pounds, carbonate of magnesia 23 pounds, and sulphate baryta 8 pounds.

LIMEITE.—A cement that is manufactured from melted India-rubber, with the addition of 8 per cent. of tallow, with sufficient slaked lime to give it the consistency of soft paste. The addition of 20 per cent. of vermilion causes the mass to harden immediately .

MADANITE.—A binding material for smooth surfaces, such as air-pumps, etc., made of 2 parts by weight of vaseline, and 1 part India-rubber, melted. This mixture may be left for years without perceptible alteration. A low grade gum used in the same way in connection with vaseline makes an excellent insulating tape, and has also been used as a friction gum.

METALINED RUBBER.—A name used for compounds used in dental work, under a process patented by C. S. Leadbetter, Manchester, England, for strengthening the gum with a metallic fabric, woven or knit.

MOROCCOLINE.—An imitation leather made from a secret compound which presumably has India-rubber for its base. Made in various colors but chiefly as an imitation of Morocco leather. An American product.

OKONITE.—A well-known compound for insulating wires and cables. According to an English analyist, it consists of India-rubber, 49.6 per cent.; sulphur, 5.3 per cent.; lamp black, 3.2 per cent.; zinc oxide, 15.5 per cent.; litharge, 26.3 per cent.; and silica, 0.1 per cent.

PANTASOTE.—A secret compound, probably of oxidized oil, which is used for the manufacture of artificial leather coverings for furniture, bookbindings, etc.

PEDRYOID.—A rubber-like finish for cloth, made presumably of oil, in tan, brown, olive, and other colors, and used chiefly in shoe finishing.

RATHITE.—A mixture in which waste silk fibers are incorporated with India-rubber to impart resiliency and durability. About 6 ounces of silk are used with 28 pounds of rubber compound. It is employed in making tires, pump valves, packings, etc. Patented by A. I. Rath, Cheshire, England.

RUBBERIC.—Fiber blended with India-rubber in solution, stretched, and dried. Used chiefly in making rubber tires and mechanical rubber goods. Patented by William Golding, Manchester, England.

RUBBER VELVET.—Manufactured by sprinkling powdered felt

of a variety of colors over proofed cloth before vulcanization. The result is a velvet-like fabric, elastic and waterproof.

THESKELON CEMENT.—A metalic substance used for waterproofing and for certain kinds of packings. It will neither expand, contract, nor rust. It is used instead of wax for sealing purposes, and resists acids, alkalies, and grease. It is often used in place of asphaltum. It can be mixed with tar, pitch, asphaltum, and other similar ingredients, the compound possessing extraordinary adhesive power. Patented by Thomas Smith, London.

VULCANINE.—A mixture of India-rubber, asbestos, litharge, lime, and powdered zinc, to which is added a percentage of sulphur. Mentioned in a patent granted to J. E. Hopkinson, West Drayton, England.

WHALEITE.—See Woodite.

WOODITE.—A name suggested by Sir E. J. Reed for an India-rubber compound invented by Mrs. A. M. Wood. It is said to possess the elasticity of India-rubber, to be uninflammable, and not injured by salt water. It is used in making valves, packings, etc. It is claimed that it will not become sticky or soft under heat or steam pressure, and will stand hot grease and other lubricants, and neither acids, alkalies, nor wastes from oil refineries, distilleries, etc., affect it in the least. A compound for Woodite or Whaleite packing is: Asbestos fiber 38 pounds, asbestos powder 38 pounds, earth wax 6 pounds, charcoal finely ground 9 pounds, ground whalebone 20 pounds, Para rubber 80 pounds, and sulphur 5 pounds.

IV. RECLAIMED RUBBER.

RECLAIMED rubber, known also as recovered rubber, shoddy, and crumb, is produced from worn-out rubber goods. There are two methods in vogue, known respectively as the mechanical and the chemical processes. The most satisfactory reclaimed rubber is made from old rubber shoes. Where the mechanical process is followed, the rubbers are ground to a fine powder, which is run over magnets to extract the iron, and is then put through a blowing process, which separates and woolen or cotton fibers from the rubber. The rubber powder in then subjected to a high degree of heat (the process known as devulcanization), and afterwards sheeted, when it is very similar to unvulcanized compounded rubber.

The chemical process is very similar to the mechanical, except that the fiber is destroyed by means of acid solutions and quite a percentage of it is washed out with the residue of the acid after the process is finished. Special grades of reclaimed rubber are made from mechanical goods that have high grade frictions in them and also from unvulcanized scrap. Rubber is also reclaimed from ordinary mechanical goods, such as hose, belting, and packing, and for certain purposes is mixed with what is known as shoe shoddy. White scrap, from wringer rolls, tubing, druggists' sundries, and the like, is also produced. The great trouble with the white is that, on second vulcanization, it is apt to be very hard. At one time, hard rubber dust was to be found in the market and was used as a shoddy in certain grades of vulcanite. There is to-day but very little of it to be found, however, as most of the manufacturers of hard rubber goods find a use for all that they make.

The processes followed in the reclaiming of waste rubber are no longer secret. Those who are in the business of manufacturing for the trade are able to do it as a rule because they buy waste stock in very large quantities at a lower figure than a small user could, besides which, by manufacturing the goods in large quantities, they can do it more economically than it could be done in a small way. It is not exposing any trade secrets, therefore, if one briefly reviews the various processes employed.

Almost the first attempt at recovering rubber waste was that done at the Beverly Rubber Works, in Massachusetts, back in the fifties, when Hiram L. Hall boiled waste vulcanized rubber in water, after reducing it to a powder, and then sheeted it. It is a curious fact, that in one little mill in the United States to-day, the manufacturer grinds his own scrap, boils it in hot water until it is in condition to sheet, and really makes a fair article out of it.

The year after Hall's patent was granted, another was granted to Francis Bashchnagel, who paved the way for devulcanization by covering a process whereby a finely ground rubber was exposed to the action of live steam. It was not, however, until E. H. Clapp took hold of the business and discovered a process for blowing the fiber out of the finely ground rubber prior to its devulcanization that the goods began to be used to a large extent.

The next step in the progress of the art was characterized by the taking out of a great variety of patents, most of which depend upon various acids and alkalies for destroying the fiber. These patents were more than fifty in number, and were fully reviewed with their attendant processes in the famous suits brought by the Chemical Rubber Co. against The Goodyear's Metallic Rubber Shoe Co. and the Raymond Rubber Co. While it would be tedious to go into that matter, it is interesting to touch upon the important processes involved. The action of acids upon fibers, of course, had long been known; in connection with the rubber business, however, it was without doubt novel. The Hayward patent, for instance, mixed 75 pounds of sulphuric acid with 8 hogsheads of water, and in this way the fiber was weakened so that it was easily ground up with the rubber. The Faure patent called simply for the immersion of the clippings in an acid, which in disintegrating the textile matter set the India-rubber free. Hiram Hall advised the use of lime or alum to eat up the cloth, and also a solution of 1 part of sulphuric acid to 9 parts of water. Burghardt used muriatic acid for destroying the cloth fiber. The Heinzerling patent called for a treatment first with acids, and then with alkalies. It is also to be remembered that Charles Goodyear directed that crude India-rubber should be subjected to a 10 per cent. solution of sulphuric acid to eat up the bark with which the gum might be contaminated.

The Mitchell patents, the Bourn patents, and others, where an extremely dilute acid was used, and where a concentrated acid was called for, have been so thoroughly reviewed that those familiar with the rubber business know all about the processes employed.

In addition to those that are now in use, a few unusual ones may be interesting. For example, the Torstrick process, in which dilute nitric acid and fusel oil were mixed with the gum in a heated state, or passed through it in the shape of vapors, making the mass sticky, after which a small quantity of chloride of calcium was added and the gum sheeted.

Conrad Poppenhusen mixed rubber scrap with essential oils,

a little turpentine being used preferably, left the scrap until it had become soft, and then passed dry gaseous ammonia into the mass, forming a gelatinous viscid product.

C. F. E. Simond mixed 2 parts of chloride of lime with 100 parts of waste rubber, and brought it to a high degree of heat, by which the sulphur was volatilized, which took from 15 to 60 minutes and then used the rubber over.

Thomas J. Mayall mixed vegetable tar with waste rubber—exposed it to the heat of the sun, or to a gentle artificial heat, and got a soft pasty mass that he was able to work with crude rubber. He also invented a process for sprinkling the finely ground rubber with camphine and setting the mass afire in a partially covered vessel, his claim being that if the fire was stopped at a certain point, a tough viscid mass was the result, which contained neither sulphur nor fiber, and could be reworked like unvulcanized rubber.

Beylikgy exposed vulcanized rubber for a number of days to a temperature of 250° F., after which he claimed that it became an adhesive mass, insoluble in alcohol, partially soluble in ether, and wholly soluble in benzole. He called this caoutchoucite and claimed that it could be vulcanized with the addition of sulphur at a lower temperature than ordinary crude rubber.

McCartney, of Glasgow, mixed vulcanized rubber with naphtha and a little acetic acid. He also added camphor, and by the action of heat produced in reality a rubber paint.

These are but a few of the many processes that have been employed, and this information, in connection with the rubber superintendent's knowledge of his particular problem, may in some cases enable him to reduce intractable and valueless wastes to a condition where it can be used in the factory.

The following are the principal grades of reclaimed rubber now on the market, a few manufacturers using copyrighted or distinctive names:

"Eureka" rubber.—The highest grade of black reclaimed rubber.

"Atalanta."—A name for a good grade of reclaimed rubber of Eureopean manufacture.

"Pongo."—A name for American reclaimed rubbers when they are sold in the European market.

"Excelsior."—A high grade of reclaimed rubber said to be made largely of unvulcanized clippings.

"Acme."—A fine grade of American reclaimed rubber.

"White Extract."—A good grade of white reclaimed rubber sold in the American market under various names.

The reclaimed rubber that is made of old shoes is usually marketed in two grades only, which are "Standard" and "XXX.," the difference being well expressed by the price, which differs a cent to the pound.

Special grades made of tires, inner tubes, air-brake hose, etc., are marketed, but are usually named for the kind of scrap from which they are taken.

V. CELLULOID AND CELLULOSE PRODUCTS.

Celluloid is made in the main from camphor and nitro-cellulose in alcohol, ether being sometimes employed as an additional solvent. The paste formed in this way is warmed gently, and then rolled out into thin sheets. The product is a brittle horny mass, consisting of a chemical, or at least an intimate, mixture of camphor and pyroxyline. A great variety of coloring matters may be added to it, and it is susceptible to manipulation and processes whereby it has been made quite flexible and practically incombustible. Crude celluloid has a specific gravity varying between 1.25 and 1.45, and has a strong odor of camphor.

Cellulose is a pure substance forming the cellular tissue of plants. In the arts use is made generally of cotton or filter paper which has been treated with acids to dissolve out impurities, and forms a basis for the manufacture of celluloid, gun cotton, pyroxoline, and xylonite. On analysis it shows: Carbon 44.44, hydrogen 6.18, oxygen 49.38. It is dissolved in sulphuric acid, and is converted into dextrine, and, by prolonging the action, into glucose. So far it has not been used largely in rubber compounding, but both alone and in connection with various other ingredients has been applied as a waterproofing. It is the basis of certain Swiss puncture fluids.

Gun Cotton.—Prepared by treating cotton wool with a mixture of strong sulphuric and nitric acids, or nitrate of potash may be substituted for nitric acid. After treatment with acid the gun cotton is rinsed carefully in cold running water, and then dried by pressure or by exposure to the air. All acid should be removed to prevent danger of explosion. Gun Cotton has been used to render fabrics waterproof, for varnishing India-rubber to render it impervious to gases, and in insulation work. Alexander Parkes, as far back as 1855, used a solution of Gun Cotton with gums or resins to take the place of compounds of India-rubber. He rendered Gun Cotton less inflammable by using biphosphate of ammonia, magnesia, talc, alum, or similar substances. As a good solvent for Gun Cotton, he distilled in 1 gallon of naphtha from 2 to 6 pounds of chloride of calcium. Charles Macintosh used as a solvent equal parts of wood spirit and coal tar naphtha.

NITRO-CELLULOSE.—This is produced by the action upon cellulose of nitric acid or a mixture of nitric and sulphuric acids. According to the length of time the acid is allowed to act, the resulting nitro-cellulose contains either 53.7, 43.6, 36.7, or 28 per cent. of nitric acid (nitric-anhydride). Gun cotton is usually a mixture containing a higher percentages while Pyroxyline—or as it is sometimes called, soluble cotton—is a mixture of a lower compounds. The solution of pyroxyline in a mixture of alcohol and ether is called Collodion.

PYROXYLINE.—A species of gun cotton less explosive in its qualities, prepared from cellulose by means of nitro-sulphuric acid. Its solution in a mixture of ether and alcohol is called Collodion.

XYLONITE.—See Celluloid.

CHAPTER VII.

RESINS, BALSAMS, GUMS, EARTH WAXES, AND GUM-LIKE SUBSTANCES USED IN RUBBER COMPOUNDING.

A GREAT variety of vegetable, mineral, and animal resins and waxes find uses in admixture with India-rubber and Gutta-percha. Their important uses are to render compounds adhesive, as in frictions, to assist in insulation, to add luster, and to modify the texture of the vulcanized compound. Many gums, like many earths, lend special virtues which they possess to rubber compounds. The more important of these materials, and those most generally used, are described in the following pages.

ADAMANTA RESIN.—An imitation copal, manufactured from common resin by a special hardening process. It is not soluble in alcohol or benzine, but completely so in boiling turpentine. It is free from acids and alkalies, and has the same melting point as Zanzibar copal. It is used rarely in rubber shoe varnish, and often in cheap frictions in mechanical lines, being moistened with resin oil to increase its adhesiveness.

AMBER.—A fossil resin chiefly found in Prussia, on the shores of the Baltic sea; it occurs also in Sicily and sometimes in the United States. It is the hardest and heaviest of the resins. Its specific gravity is about 1.07. By distillation a yellow oil—oleum succini or oil of amber—is obtained, and a yellow resin remains in the still. Amber varies in color from light yellow to a deep brownish red. It is insoluble in almost all of the ordinary solvents. When heated above its melting point, however, it becomes partly decomposed, and is then soluble in oil of turpentine and alcohol. It makes a very fine transparent varnish, which is used on negatives in photographing. It is used in cements for fastening lineoleum and rubber tiling to decks, and is also mentioned in the formulas for certain patented gums.

ASPHALT is undoubtedly an oxidized residue from evaporated petroleum. This name is applied usually to the solid bitumen, the liquid being called mineral tar, and sometimes maltha. It is chiefly made up of hydrocarbons, but contains a certain amount of sulphur and nitrogenous bodies. It is known also as natural pitch,

Jews' pitch, asphaltum, bitumen, etc. It is a black hard substance which, when freshly broken, shows shining surfaces that are always correspondingly rounding and hollowing. It is insoluble in water and alcohol, but dissolves in benzine, acetone, and carbon disulphide. Is used in rubber compounding in place of coal tar, and in insulating compositions, and in certain substitutes like Kerite. Commercially there are two grades, known as "lake pitch" and "land pitch," of which the latter is the harder.

In solution it is used sometimes to protect rubber goods that are exposed to the destructive influence of brine. A little Asphalt is also said to increase the elasticity of hard rubber. Asphalt mixed with resin and oil of tar forms a low grade artificial Gutta-percha. It is added to "Cooley's artificial leather" to harden it and enable it to resist heat. It is also the basis of one type of marine glue.

ARTIFICIAL ASPHALT.—This is made by heating sulphur and resin together to about 250° C., where the reaction takes place, attended by the evolution of sulphuret and hydrogen, and leaving an almost black, pitchy substance resembling asphalt. It is insoluble in alcohol, but dissolves readily in benzine.

AUVERGNE BITUMEN.—A species of natural asphalt found in the province of Auvergne, France. It is similar to Trinidad asphalt, but is impure, containing clay, silica, magnesia, iron, and traces of arsenic. (See Asphalt.)

BALSAM.—This term is given to oleo resins which are soft at ordinary temperatures, and are really a mixture of such a resin and the essential oil of the plant from which they exude, such as benzoin, tolu, etc.

BALSAM OF STORAX.—Produced from the inner bark of a tree of the genus *Storax,* in Asia Minor. Commercially it is a soft, coarse, dark colored powder, or, more commonly, a semi-fluid, adhesive substance, brown outside, greenish gray inside. The sweet gum of the southern United States is allied to the Eastern drug, and was formerly much used in chewing gum. Used in general cements, being particularly good in leather cements; also for glass, stone, and earthenware cements.

BALSAM OF SULPHUR.—A solution of sulphur in boiling volatile or olive oil. Used in certain rubber compounds as a vulcanizing agent and a protection against blooming.

Beeswax is obtained from the comb built by honey bees. The crude wax is yellow and soft, with a granular fracture. Its specific gravity varies between .965 and .969, its melting point being between 140° and 144° F. It is often adulterated by water, by white mineral powders, and by cheaper substances, such as vegetable wax, paraffine, etc. White wax is that which has been exposed to the sun or to the moderate action of nitric or chromic acid, thereby being bleached. It is sometimes used with rubber in medicinal plasters. Ordinary beeswax is largely used in the valuable hard rubber compounds known as the Kiel compounds. Sheet beeswax is often used in the work of vulcanite pattern making. It is also used in processes for making fabrics water-repellent, the other ingredients being aluminum, resin, soap, wax, and silicate of soda. With Gutta-percha it is an ingredient in shoemakers' wax, and also in certain proofing compounds. Hancock used it in a Gutta-percha compound for a soft effect. In a hard rubber compound made up of India-rubber, sulphur, oil, and pumice stone, it is said to be acid proof.

Birch-bark Tar.—A peculiar tar obtained during the distillation of birch-bark for oil, being probably the same as Russian Jackten extract. Used in the manufacture of certain rubber substitutes.

Bitumen.—The term applied to a body made up of several hydrocarbons. It resembles Trinidad asphalt and is of the same nature. Its specific gravity is from 1.073 to 1.160. Artificially it is prepared from shales, mineral asphalt, etc. It is used as a source of paraffine. The West Indian product is known as Chapapote. A solution is made from it in which the tapes are soaked that are used for covering wire that has been insulated with India-rubber. Bitumen has been utilized by what is known as the calender process, which is a partial vulcanization, rendering it valuable as an insulator.

Black Pitch.—Is the residue left after the oils of tar have been distilled from that body. Used in weather proofing work.

British Gum.—See Dextrine.

Burgundy Pitch.—Is obtained from the hardened juice or sap which concretes upon the bark of the Norway spruce. As imported it is often quite impure and should be melted and strained before being used. It is almost entirely soluble in glacial acetic acid or boiling alcohol, and somewhat in cold alcohol. When pure it is hard and brittle, with a shining fracture, reddish or yellowish-brown, aromatic. It is much used in cements, in electric tape, and in the manufacture of porous plasters. Common resin is often melted and mixed with fats and water, forming a gum that much resembles Burgundy Pitch.

Burmite Amber.—Found in Burma, but quite inferior in quality. It is a little harder than amber proper, is easily cut, takes an excellent polish, but has less variety of color. (See Amber.)

Button Lac.—See Shellac.

Canada Balsam.—Sometimes called Canada turpentine. It is derived from the *Abies balsamea*. It is a yellowish or greenish transparent liquid, completely soluble in ether, chloroform, or benzol. It is sometimes called Balsam of Fir, but it does not really belong to the balsams, being a true turpentine. Strasburg turpentine is sometimes substituted for it commercially.. It is used in certain compounds to prevent sulphur from efflorescing. With paraffine, beeswax, and coloring matters, it is used for insulating colored yarns that are used for anunciator and similar wires, and it was also used by Duncan in Gutta-percha cements for leather.

Candle Tar.—The residual products from the distillation of animal fats, oils, etc., are known as candle tar. This product is sometimes soft and ropy, and at other times quite hard. Mixed with sulphur, it is said to produce a compound having some of the elasticity and other desirable qualities of vulcanized India-rubber.

Casein (also called Caseum) is one of the chief constituents of milk, being that part which forms the curd of sour milk, and is familiar in the form of cheese. A similar substance, prepared from peas, beans, lentils, and the like, is called vegetable casein. It is used in shower-proofing after a German formula in connection with soda, lime, and acetate of alumina; also, in cements of which Gutta-percha is the base, for joining small particles of leather, shavings, etc.

Carnauba Wax is found in Brazil, where it forms as a coating on the leaves of a certain palm (the *Corypha cerifera*), and is removed by pounding and shaking. It is very hard and is of a greenish or grayish color. Its specific gravity is about 0.995, it is odorless, and melts at 185° F. It dissolves completely in boiling alcohol, and is used on insulated wire as a finish, and in the manufacture of wax varnishes.

CARN GUM.—Used instead of ozocerite as a finish for tape or braids that cover insulated wire. (See Carnauba Wax.)

CERAMYL.—A material used in the finishing process in the manufacture of elastic web. Its use is to make the web stronger, and in a measure to act as a size, causing it to lie flat. It is also said to add strength to it. By the application of heat, ceramyl, which comes in the form of a semi-solid, is reduced to a liquid. In English practice this is said to have driven out the use of glue in the dressing of elastic webs. Ceramyl is manufactured in England.

CERASIN, also spelled Ceresine, is of a butter yellow color, odorless, and has a specific gravity of .918 to .922. It is used chiefly in covering anunciator wires where the object is to preserve the colors of the yarns in the braiding. (See Ozocerite.)

CHERRY GUM.—A pale yellow or red brown gum, coming from the bark of old cherry trees. It contains 35 per cent. of cerasine, 52 parts of arabicum, and 1 to 3 per cent. of ash. This gum is chiefly used in the manufacture and finishing of fine felt hats. The gums on the market are of two qualities, the German, which is the best, and the Italian. It is used in insulating instead of purified ozocerite, in certain cases where a little more adhesiveness is required.

COAL TAR.—See Tar.

COLOPHANE.—See Rosin.

COLOPHONY.—See Rosin.

COORONGITE.—The name given to a rubber-like mass found in Coorong, South Australia. Some place it among the fossil resins. Coorongite is not soluble in the ordinary solvents used in rubber work, but, after mixing with India-rubber, it can be put in solution. According to Forster, it vulcanizes somewhat as India-rubber does. (See Pseudo Rubbers.)

DEXTRINE is a sort of intermediate product between dextrose and starch. It is soluble in cold water, and is much used as a substitute for gum arabic in mucilage, as it has strong adhesive properties. Cooley combined it with a little Gutta-percha, resin oil, and earthy matters in the production of what he called artificial leather. It is used also in a mixture with plaster of paris, making a tough surface mold for small experimental rubber work.

DEXTROSE is obtained from starch generally, and is crystallized glucose. It is soluble in water, and has many commercial uses. For example, it was used by Hancock as a sizing for cloth on which was spread rubber in solution, the Dextrose being there in order to keep the rubber from sticking to the cloth. In other words, this was a sort of cheap calendering process.

EARTH WAX.—See Mineral Wax.

ELATERITE is also known as elastic bitumen or mineral caoutchouc. It appears naturally in soft, flexible masses of a brownish black colors somewhat resembling India-rubber. It is composed of 85.5 per cent of carbon, and 13.3 per cent. of hydrogen. In its physical characteristics, Elaterite is found in infinite variety. It is sometimes elastic and so soft as to adhere to the fingers, and sometimes brittle and hard. One kind of it, when fresh cut, resembles fine cork both in texture and color, and will rub out pencil marks. Its elasticity is due to its cellular texture, and to the moisture with which it combines. It is used to a certain extent in insulating compounds, but is intractable and so far shows no special features of value above other minerals of the same series. A few years ago a company was formed in Colorado which claimed to be able to make many kinds of rubber goods from this product, alone, but little has been heard of the plan of late. (See Gilsonite.)

ELASTIC GLUE is used with India-rubber and Gutta-percha in shoemakers' cements. (See Substitutes.)

FRENCH ASPHALTE.—See Auvergne Bitumen.

FICHTELIT.—Occurs in a peat bed near Redmitz in the Fichtelgebirge in Germany, and also in fossil pines in the form of scales or flat needles. It has also been met with in Franzenbad and in Denmark. A hydrocarbon little known, though mentioned in certain patented rubber compounds.

FISH GLUE.—Made by boiling the heads, fins, and tails of fish by high heat. It is generally made into a liquid glue by a treatment with acetic or hydrochloric acid, whereby its property of gelatinizing is lost. It would have a disagreeable odor were it not for the fact that that is destroyed by adding creosote oi oil of sassafras or something of that kind. Fish Glue is used in a cement for cured rubber, in connection with Gutta-percha and rubber dissolved in bisulphide of carbon. (See Glue.)

GARNET LAC.—See Shellac.

GILSONITE.—A hydrocarbon valued for its elasticity. One of the purest of crude bitumens, it is mined in the Uncompahgre Indian reservation, Utah, United States. It is a black, tarry-looking substance of brilliant luster. It is used for varnish making, in paints, and for insulation, either with or without rubber, one well-known compound consisting of rubber, linseed oil, and Gilsonite.

GLUCOSE.—The commercial form is prepared from starch usually, as that is the cheapest raw material. The starch paste being boiled with mineral acids, dextrose, maltose, and dextrine are produced. Glucose in this country is made entirely of corn-starch; in Europe, however, sago starch, rice, and potato starch are used. It is neutral, and both odorless and colorless. It is really a kind of sugar that is with difficulty crystalizable, and it is also called grape sugar. It occurs in commerce either as a thick, sweet, heavy liquid, or as a white solid mass. It is used with rubber glue, sugar, whiting, and glycerine in making bookbinders' cements, and in making puncture fluids for pneumatic tires.

GLUE.—An impure form of gelatine obtained from the horns, hoofs, skins, and bones of animals. Glue of good quality should be bright brown or brown yellow in color, free from specks, glossy, perfectly clear, hard, and brittle, should not become damp by exposure to the air, and should snap or break sharply when being bent, the fracture showing a glassy, shining appearance. Used in bookbinders' cements, in cheap frictions, and in cheap horse-cover compounds with rubber. A size made of glue was used by Brockedon to protect fabrics that come in contact with the liquid used in cold curing. This was afterwards dissolved off by an alkaline solution.

GLUGLOSS GELATINE.—A gelatinous product used largely in Amercia in waterproofing fabrics. It is dissolved in hot water to use, and makes an excellent waterproof sizing. A mixture of glycerine with it increases its elasticity. It combines readily with glue, dextrine, or any such products, and develops considerable adhesiveness.

Gluten.—A vegetable substance obtained from wheat and other grains. Treated with tannic acid, it is used as a substitute for Gutta-percha under a formula by Johnson, who says the product can be vulcanized. Another formula calls for its mixture with oil and sulphur, as a substitute for Gutta-percha. In cements it is the basis of one for uniting leather scraps, and is used with a little Gutta-percha.

Gum Anime is a South American fossil resin similar to copal. It occurs in small irregular pieces of a pale yellow color. Has a high melting point, and its specific gravity is 1.028 to 1.072. Mixed with rubber and earthy matters and dissolved in turpentine, it formed one of the early compounds for clothing.

Gum Arabic is an exudation from a species of *Acacia*. It is made up of clear, or semi-transparent fragments, hard and brittle, breaking with a shining fracture. It is inodorous and feebly sweetish to the taste. Its specific gravity is 1.31 to 1.52, for dried gum. It comes from Africa and is known also as Acacia and Gum Senegal. It dissolves in hot or cold water. It is used in connection with plaster of paris in making a tougher surface mold for small and experimental rubber work. Enough gum is added to make the mixing solution about the thickness of a thin syrup. It is largely used in cements. It is also used in certain shower-proof compounds, and in paste blackings made of caoutchouc oil, vinegar, molasses, and boneblack.

Gum Ammoniacum.—Exclusively obtained from Persia as tears, or aggregated masses, of a peculiar smell and a taste slightly sweetish, bitter, and somewhat acrid. Its specific gravity is 1.207. Used in solutions for pressed leather cuttings and fibrous wastes. Ten parts of this gum mixed with 20 or 25 parts of Gutta-percha form a cement possessing both elasticity and solidity, and is thoroughly waterproof, used for filling cracks in horses' hoofs. Also used with Gutta-percha, boiled linseed oil, and caseum or casein, for sticking together small particles of any dry matter in the production of artificial leather.

Gum Benzoin.—Occurs in lumps of yellowish brown tears, stuck together and more or less mottled from the white inside the tears. Its specific gravity is from 1.063 to 1.092. Of an agreeable balsamic odor and very little taste, but irritating when chewed for some time. Used in linseed oil proofings, presumably to kill odor; also in certain Gutta-percha and India-rubber compounds for disguising the odors. Four per cent. of the weight of the mass is said to be sufficient to make the odor an agreeable one. According to Forster, a little of it mixed with Gutta-percha greatly improves the quality.

GUM ASPHALTUM.—Refined natural bitumen, also called litho-carbon. Is found in Texas and at one time was exploited as a substitute for rubber. (See Litho-Carbon.)

GUM CAMPHOR.—The white transparent substance known by this name is obtained from Japan and the island of Formosa. It is really an oxygenated essential oil. Its specific gravity is 0.985. Sparingly soluble in water, and very soluble in alcohol, ether, acetic acid, and hydrocarbons or volatile oils. Is largely used in the manufacture of celluloid. Gum Camphor is also used in compounds of the substitute order like Textiloid, Kerite, etc. Was also the basis of several remarkable compounds known as Heevenoid (which see).

GUM COPAL.—Hard Copal is a fossil resin obtained from the East Indies, South America, and the eastern and western coasts of Africa. It occurs commercially in roundish, irregular pieces, having a specific gravity of 1.045 to 1.139. It is insoluble in alcohol, partially soluble in ether, and slightly so in oil of turpentine. Soft Copal is obtained from living trees in New Zealand, the Philippine islands, Java, and Sumatra. Used with shellac, asphaltum, and arsenate of potash for waterproofing leather; also in cements, in proofing compounds, and in varnishes in connection with India-rubber, lead, alum, and other ingredients dissolved in spirits of turpentine.

GUM DAMMAR is derived from the Amboyna pine, growing in the Malay peninsula, Sumatra, and Borneo. The resin exudes in tears and is collected after it has dried. It makes a very transparent varnish, the gum being soluble in benzine, essential oils, and to a certain extent in alcohol. Used in artificial leather compounds, and with rubber, asphalt, and fish oil for waterproofing leather. It is quite largely used in rubber cements.

GUM ELEMI comes from the Philippine islands, and is a rosin obtained from certain trees there. It varies from white to gray in color, and is quite soft and very tough. Alcohol and other solvents readily dissolve it, and its office usually is to give toughness to varnishes in which are harder resins. Used in connection with India-rubber and benzine in the production of puncture fluids. (See Manila Gum.)

GUM EUPHORBIUM appears in the market in the shape of tears of irregular shape, varying in size from a small pea to 1½ inches in length. Of a dirty gray or yellowish color, and very largely mixed with impurities. Must not be confused with Gum Euphorbia (which see.)

GUM FRANKINCENSE.—Also called Olibanum (which see.)

GUM GAMBOGE.—The best is found in commerce in cylindrical rolls of a dull orange red color. Another form is that of lumps or cakes. Its powder is bright yellow and its taste very acrid, but it has no smell. It is derived from a tree which is a native of Cochin China and Siam. Is used chiefly as a pigment. It is the basis of a general cement in which is also found rubber, alum, and burnt sugar, and in another is used with rubber, white lead, gum benzoin, alum, sugar, and sulphur, for cementing vulcanized rubber.

GUM LINI.—A gum made from linseed, often used as a substitute for gum arabic. The seeds are first boiled in water for an hour, the resulting thick mass filtered, and then treated with twice its volume of 90 per cent. spirits of wine. A flocculent white precipitate separates, from which the dilute spirit can readily be decanted. The gum is clear, grey brown, fragile, and dissolves in water. Two grams in 30 grams of oil is almost identical with an emulsion of gum arabic. In connection with coloring matters is the basis for the Knowlton patented waterproofing process.

GUM TRAGACANTH is an exudation which comes in the form of translucent plates of a dull white, which water swells and partly dissolves. It is often used in mucilage in place of gum arabic. The gum comes from the Levant from the *Astragalus gummifer.* Has been used in connection with Gutta-percha for making dental plates that are soft and adhesive to the membranes and that will not rot or deteriorate.

GUM LAC.—See Shellac.

GUM TRAGASOL.—This is a gum produced from the kernels of the *Ceratonia siliqua.* The use of this gum as a solvent for India-rubber, Gutta-percha, or celluloid has been patented in England. A mixture of 25 parts of dissolved India-rubber, 75 parts of strong gum solution, with the addition of 1 part of carbolic acid to 500 parts of the mixture, makes a cement for wood, and a preservative paint against insects and vermin.

GUM JUNIPER is the gum known as sandarac, obtained from an evergreen growing in northern Africa. It occurs in small, light-colored grains, with a slightly bitter taste. It is soluble in turpentine oil and alcohol. Is used as an assistant in making peroxide substitutes. Mixed with rubber and earthy matters and dissolved in turpentine, it was one of the early compounds for clothing.

GUM OLIBANUM.—The frankincense of the ancients, obtained chiefly from Asia and Africa. It occurs in yellowish, somewhat translucent tears, with a balsam-like resinous smell, and an acrid aromatic taste. Sometimes called Gum Thus. It is largely used in the manufacture of porous plasters.

GUM THUS.—A name for gum turpentine, and rarely for olibanum. Used with rubber and Japan for waterproofing leather.

GUM TURPENTINE.—Turpentine hardened by exposure to the air. (See Turpentine.)

HELENITE.—Another name for fossil rubber or Elaterite (which see.)

ISINGLASS.—A substance prepared from the swimming bladders of certain fish. It is white and glistening, occurring in fibers or threads. The best is known as Russian, and comes from Astrachan. Its specific gravity is 1.2. On boiling isinglass it is converted into a very pure form of glue. Isinglass is used in quick drying cements with India-rubber, chloroform being the solvent.

IDRIALIN (IDRIALIT).—A rare hydrocarbon found in Idria, a province of Austra, where it occurs with hepatic cinnabar. A similar body is obtained in the distillation of amber. Its specific gravity is 1.4 to 1.6. Mentioned in certain rubber formulas to assist the insulating qualities of compounds.

Kauri Gum.—An amber-like substance varying from a soft cream white to an amber color. It comes from New Zealand, and is also known as Australian dammar. The lighter colored Kauri comes from living trees, but much of the darker is a fossil resin. It is cheaper than copal and largely used in varnishes. Kauri Gum, in connection with rubber gum and pitch, is used for treating yarns used in insulated wire coverings. Parkes added it to rubber goods where the surface was to be printed upon after curing. One pound of Kauri, 8 pounds of Gutta-percha, and 1 pound of milk of sulphur formed Richard's covering for insulated wire.

Lac.—See Shellac.

Litho-Carbon.—A kind of asphalt large deposits of which are found in the state of Texas. It was at one time thought that it would supersede India-rubber, and a company was formed with the idea of manufacturing goods from it. This was in 1892, and India-rubber is still used. The chemical composition of Litho-Carbon is 88.23 carbon, 11.59 hydrogen, .06 oxygen, a trace of sulphur. Litho-Carbon is jet black in color, is flexible at ordinary temperatures, and is quite tough. Its specific gravity is about 1.028. It is said to be soluble in naphtha, benzol, bisulphide of carbon, etc. It will stand a temperature of 600° F., without giving off its associate products. It resists alkalies and acids, with the exception of concentrated nitric and sulphuric acids. Its manufacture was patented. Used with Gutta-percha and shellac it makes an excellent insulator.

Manila Gum.—See Gum Elemi.

Mastic.—A resin from the shores of the Mediterranean. It occurs in tears of a pale yellow, is brittle, and of a faint balsamic odor. It dissolves in acetone, turpentine oil, and alcohol, and is largely used in varnish. The residue obtained in the purifying of mineral asphalt is also called mastic. It is used in general rubber cements for joining stoneware, earthenware, leather, etc. One of special value calls for 10 parts of mastic to 1 part of India-rubber, dissolved in chloroform, and makes an excellent cement for fastening letters to glass. The gum also appears in many old fashioned compounds.

Menthol is obtained from the oil of peppermint coming from Japan and China, or from the oil of spearmint manufactured in the United States. Its melting point is about 108° to 110° F., and it is slightly soluble in water, but freely in alcohol. It is often used in medicinal plasters which have rubber for a base.

Mineral India-rubber Asphalt is the name of a material composed of refuse tar produced during the refining process of tar by sulphuric acid. It is black, like ordinary asphalt, and quite elastic. It is an excellent non-conductor of electricity, and is not assailed by acids or alkalies. In a naphtha solution, it yields a waterproof varnish for metallic objects, and is used in rubber compounding in place of asphalt.

Mineral Tallow, also called Hatchetine, is a substance found in Siberia, Germany, and Great Britain. It is an earth wax that is soft, flexible, and runs from yellow to yellowish white. It has no smell, and melts at from 115° to 170° F. It is composed of 14 hydrogen and 86 carbon. Mineral Tallow is used sometimes in place of earth waxes in insulated wire work, and has been used in paste blackings in connection with India-rubber.

Mineral Wax.—A term applied to several waxy-looking hydrocarbons found as mineral deposits, such as neft gil (naphtadil), ozocerite, and earth wax. It is found in Austria, and in the southern part of Russia, on the shores of the Caspian sea. In the United States it occurs largely in Texas and Utah. Used chiefly in insulating compounds. (See Ozocerite.)

Myrrh exudes from the bark of a tree which grows in Arabia, in yellow drops that are quite oily at first, but which thicken and become hard and of a dark color. It appears in commerce in either grains, or tears, or in pieces of various sizes and irregular form, the color being red, reddish brown, or yellow. Its taste is bitter and aromatic, and its smell balsamic. The best gum is known as Turkey Myrrh. It is used with rubber, sulphur, and salycilic acid in complexion masks.

Natural Pitch is the name given to such kinds of pitch as are not manufactured, such as asphalt, bitumen, etc.—that is, pitch of a mineral origin, except that from coal or shale. (See Asphalt.)

Oleo Resins.—A resin that contains a certain amount of the essential oil of the plant from which it exudes is so called. Chief among the Oleo Resins are certain which have a pungent taste and a peculiar, and often a pleasant odor, known as balsams.

OZOCERITE.—A waxy hydrocarbon occurring in Austria, southern Russia, and the United States. It is also known as earth wax. Its specific gravity is 0.9 to 0.95, and it is about as hard as talc. Chemically, it consists of hydrogen 13.75 and carbon 86.25, while its melting point extends from 140° F. to 170°F. It is often found adulterated with asphalt and sometimes with Burgundy pitch. Purified Ozocerite is known as ceresine. To make this, the crude material is treated with fuming sulphuric acid, and then filtered through charcoal. Thus prepared it is of a pale yellow color, the melting point ranging from 61° to 78° C. It has almost wholly driven out Stockholm tar as a protection for wires insulated with Gutta-percha, when placed under ground. It improves the insulation, but in spite of common belief to the contrary, does not preserve textile fabrics. The best compound for the protection of the insulation on wire consists of 3 parts of Ozocerite to 1 part of Stockholm tar. It is an insulator of high quality, and while it is in some ways intractable, its wax-like nature allows it to combine with other insulators or with textiles. It is also used as a water-repellent in fabrics, the gum being volatilized by heat, and the fumes passed through the cloth. As a surface covering for tapes or braid, it is often employed and is better than other gums, as it takes a fine polish from the polishing machine. The basis of Henley's system of curing India-rubber core is melted Ozocerite, which is used under pressure to remove all the moisture, being afterward heated in hot Ozocerite, which stops up the pores. Ozocerite, mixed with India-rubber, is also the basis of the India-rubber compound called nigrite. It mixes, however, with difficulty with India-rubber, which is an objection to many proposed uses of it. It also has a mildly deterious effect on it.

OZOCERINE is a vaseline-like substance prepared from ozocerite. There is also prepared from crude ozocerite a valuable black wax which, when fused with India-rubber, makes an excellent electric insulating material. This wax was recognized by a lecturer before the Society of Chemical Industry as the basis of the insulation known as Okonite.

PARAFFINE.—A white waxy-looking body obtained from certain tars by distillation. It is tasteless, inodorous, harder than tallow, but softer than wax. Its specific gravity is .877. It is also obtained from ozocerite or earth wax. Its melting point varies with the source it is obtained from. It is insoluble in water and nearly so in boiling alcohol, but soluble in ether, oil of turpentine, oil of olives, benzol, and bisulphide of carbon. It is usually very free from water, and not liable to absorb it. It has been used as a-waterproofing mixture and is a good insulator. A very widely diffused bit of newspaper advice has been that to preserve rubber goods they should be dipped in a bath of melted paraffine and dried then in a hot room. It has not been proved to be of any advantage, however. Experts in the rubber trade claim that such a course would seriously injure the elasticity and life of the rubber. When gossamer clothing was manufactured in large quantities, the surface of the goods before solarization was covered with a thin coat of paraffine, which gave it a peculiar shade until the solarization was completed, when all traces of the paraffine seemed to disappear. The insulating capacity of rubber to which paraffine has been added is quite remarkable, but at the same time it lessens the hardness of the rubber to a marked degree. Rubber dissolved in Paraffine wax forms a curious compound which has been used in insulation. Paraffine is used in the artificial gums like Parkesine and insulite; also with cottonseed oil and resin for cheap Brattice cloth, and in cheap proofing compounds. It is not a great favorite as an insulator, as it shrinks in cooling, causing cracks. Paraffine tapes are also easily destroyed through the presence of free acid. It was formerly used largely in covering anunciator wires, but as it was found to absorb and retain water, its use was given up, and its place taken by a compound of Paraffine, ceresin, and resin.

PITCH is the black residue that remains after the distilling of wood tar. Varieties are also obtained from coal tar and from bone tar. Wood pitch, however, has a toughness which the others do not possess. Pitch was used very early in considerable quantities in hard-rubber compounds. Goodyear, for example, used considerable of it in hard compounds for coating metal, the rest of the compound consisting chiefly of rubber and sulphur. It is almost the only organic substance which largely increases the resiliency of India-rubber. It is largely used in cements, and also in many rubber compounds. Equal parts of pitch and Gutta-percha make a tire cement for fastening to the rims, known as "Davy's Universal Cement." It is used with Gutta-percha in shoemakers' wax, and also in certain proofing compounds. Wood cements made of Gutta-percha as a rule contain a certain amount of Pitch. It is also used in the manufacture of Fenton's artificial rubber.

RESINS.—The term given to a number of complex bodies, generally the hardened exudation of sap from trees. Chemically a resin is the substance obtained by the gradual oxidation of an essential oil. The specific gravity ranges between 1.02 and 1.2. Resins are divided as a rule into three classes—hard, soft, and gum resins. The former at ordinary temperatures are solid and quite brittle. They contain little or no essential oil, and are easily pulverized. Shellac and sandarac are good examples of this kind, and soft resins are usually called balsams, and are either semi-fluid, or soft enough to be molded by hand. They are really mixtures of hard resins, and the essential oils found in the plant from which they come. On exposure to the air they become in time hard resins. Of this class are balsam of storax, tolu balsam, etc. Gum resins are the solidified milky juices of certain plants. They consist of a mixture of resins, essential oils, and a considerable proportion of gum. These are, for example, gum euphorbium, galbanum, and to this class also belong India-rubber and Gutta-percha. Most of the fossil gums, such as copal, are resins whose physical characteristics have been changed by their having been buried for a long time in the earth. These fossil resins are counterfeited to an extent by treating ordinary resin with lime which raises its melting point considerably.

RETINITE.—Also known as Retin Asphalt. It is a fossil resin found in brown coal. It is found in roundish masses of a yellow brown or reddish color, is quite inflammable and readily dissolves in alcohol. At present it is somewhat rare, but if it ever should become common, it would undoubtedly find a place in rubber compounding. Its specific gravity is 1.07 to 1.35.

ROSIN is made from common turpentine, which is distilled in water yielding nearly one-fourth its weight of essential oil, the residue in the retort consisting of common rosin. Rosin was also very generally called colophony, a name now practically obsolete. There are two varieties of rosin in common use, the brown and the white. The first named is brittle, solid, and of an amber color, and comes from the Norway spruce fir. The white rosin is obtained from the pine and is known as galipot. Rosin dissolves very freely in alkaline solutions, which allows of its use in soaps. Its specific gravity is 1.08. There are three grades commonly on the market, which are called virgin, yellow dip, and hard. It is used in a great variety of rubber compounds, its chief uses being in frictions, dry heat varnishes, cements, and puncture fluids. Almost all lines of rubber manufacture use a certain amount of it at times. Only a small proportion of it can be used in rubber compounding, its office being usually that of the sticker. A large amount of it induces surface cracking, and often a decided blooming of the sulphur. It is also used in waterproof solutions in conjunction with spermaceti, India-rubber, and paraffine wax. Mixed with boiling oil, it has been applied to Gutta-percha articles to give them a Japan-like luster, and is also important in Gutta-percha glue, which is compounded of Gutta-percha, powdered glass, litharge, and Rosin. A very large use for it is in the rubber channel cements that are sold to leather shoe manufacturers.

SANDARAC.—Also known as Gum Juniper (which see.)

SEEDLAC.—See Shellac.

SHELLAC, STICKLAC, SEEDLAC, GUMLAC.—All these are different names for the same thing or different stages of its preparation. It is the exudation formed on several sorts of trees growing in the East Indies, but is chiefly produced from the banyan tree, the exudation coming from a scale shaped insect known as the *Coccus lacca,* the female fixing herself to the bark and exuding the resinous substance from her body. In addition to the East Indian product there is what is known as Mexican lac, which exudes from the *Croton draco.* Sticklac is the resin as taken from the tree. Sedlac consists of fragments broken from the twigs and partly exhausted by water. Shellac is prepared by melting Stick or Seedlac, straining, and pouring upon a flat surface to harden. It is then washed, dried, melted, roughly refined, and sent to market, or it is poured into molds to harden and is known as Button or Garnet lac. The specific gravity of Lac is about 1.139. It is partially soluble in alcohol, turpentine, chloroform, and ether, and is completely soluble in caustic alkalies and borax solutions. Shellac was formerly used very generally in rubber manufacture in surface goods, and particularly in solarized goods in small proportions. It has a specific use to-day in the production of water varnishes for surface goods. It is also a constituent in the production of certain compounds in hard rubber, and particularly the semi-hard varieties, being used to the extent of 20 per cent. of the amount of gum. Although quite brittle, it seems to impart a certain elasticity to the product. The maximum use of Shellac in a hard-rubber compound, according to Hoffer, is 88 parts of India-rubber, 50 parts of Shellac, 12 parts of sulphur. It is also used in certain of the Jenkins patented packings to the extent of 10 to 25 per cent. of the amount of rubber, where it is said to preserve the compound from the effects of coal oil, steam, or hot water. It is also used in many cements both with and without India-rubber, one formula for marine glue being: 20 parts of shellac, 12 parts of benzol, and 1 part of India-rubber mixed with heat. Dissolved in 10 parts of strong aqua-ammonia, it forms a varnish for rubber goods, and is also used as a solution for revarnishing old rubber shoes. Used with carburet of iron and bisulphide of mercury as a cement for card clothing, with rubber and Gutta-percha for attaching shoes to horses, in English "ale cement," and in certain proofing compounds.

SIZE.—A weak solution of glue, sometimes used in showerproof compounds and cements. The name Size is also often applied to any thin viscous substance, as for instance, gilders' varnish. In rubber practice, however, the glue Size is what is ordinarily employed. It is also used in preparing a perfectly smooth cloth upon which rubber is to be calendered, and from which it is stripped before the making up. (See Glue and Gelatine.)

SPRUCE GUM is used with chicle in the production of chewing gums. Melted spruce gum or rosin is known as Burgundy pitch (which see.)

STEARINE.—A white waxy-looking body obtained from fats.—chiefly tallow and palm oil. When made from tallow it is called pressed tallow or tallow Stearine, which is the solid part obtained from the heating of suet fat and the removal of the liquid part which is oleomargarine. Tallow Stearine is very largely used in candle making, where is found saponified Stearine, distilled Stearine, and distilled grease Stearine. This latter contains considerable cholestrol and differs from commercial stearic acid or Stearine chiefly in its physical structure. Stearine is used in prooffing compounds, in rubber blackings and in compounds containing resins. It has been suggested that a small proportion of Stearine in certain rubber compounds that contain low grades of rubber which in themselves have large proportions of resin, has a decided value in preventing oxidization. Used in proofing compounds, rubber blackings, and compounds containing resins.

STEARINE PITCH.—The brown tarry residue left in the still during the process of refining tallow and fat. Used in the manufacture of certain packings that contain no rubber. Stearine Pitch is also used as a lubricant for bearings that have a tendency to heat.

STICK LAC.—See Shellac.

STOCKHOLM TAR is used in black cements of the marine glue class, and is also used in rubber compounding, its office being to assist in the mixing of dry compounds, and as a binding material for sulphur in the dry heat cure. Also used in manganese cements and in cements to fasten tiles to floors. (See Tar.)

SPERMACETI.—A peculiar fatty concrete substance obtained from the head of the sperm whale. Its specific gravity is 0.943, and it is fusible at 112° F. Insoluble in water, soluble in hot alcohol, ether, and oil of turpentine, but redeposited as the liquids cool. Was formerly used in certain waterproofing compositions.

SLUDGE OIL RESIN.—A heavy gummy residue from the waste of superphosphate factories. Has been used with rubber in making Japan varnishes.

TAR.—This substance is derived from the animal, vegetable, and mineral kingdoms. From the first, by the destructive distillation of bones, is produced what is known as "Dippel's oil"; from the second, by the distillation of pine woods, the product is known as pine tar or Stockholm tar; and from the third, by the distillation of coal, is produced coal tar. Of the three, coal tar is the most used in rubber work, its office being to help carry adulterants in dry mixing and to keep the sulphur from blooming after vulcanization. It is used chiefly in dry heat work. Goodyear discovered early that very large quantities of boiled tar could be used in connection with India-rubber and sulphur without injuring the quality of the gum, and it has been very generally used since his time.

TRINIDAD ASPHALT is obtained from the pitch lakes of the island of Trinidad. Its specific gravity is 1.2, and it is somewhat soluble in alcohol, while Persian naphtha, oil of turpentine, benzol, and benzoline readily dissolve it. (See Asphalt.)

TOLU BALSAM is derived from a tree found on the mountains of Tolu, and the banks of the Magdalena river, in Colombia. It is very similar to balsam of Peru. It sometimes appears in commerce in dry friable fragments, the newly imported gum being soft and tenacious. It has a very fragrant odor, and a medicinal and tonic effect. Tolu Balsam is used with paraffine wax and chicle in chewing gum compounds.

TURPENTINE.—This is a semi-solid resin, which comes from various species of pine as a rule. The chief commercial varieties are common turpentine, which comes from the *Pinus abies;* Venice turpentine, from the larch; Bordeaux turpentine, from the *Pinus maritima,* and Chian turpentine, from the *Pistacia lentiscus.* Of these the Venice turpentine is said to be the best. It is of a pale yellow color, transparent, has a bitter taste, but a balsamic odor. Used instead of rosin in many compounds.

VEGETABLE PITCH.—The residue left after distilling the tar made from wood of various trees. Called vegetable to distinguish it from the mineral pitch which is derived from coal. (See Pitch.)

XANTHORRHOEA GUM is somewhat similar to shellac, is abundantly produced in the Australian colonies, and sometimes used in the compounding of ebonite. Xanthorrhoea Gum is also sometimes known as gum acaroides, and is produced from the Australian grass tree.

XYLOIDIN.—An artificial gum much resembling pyroxylin obtained by the action of nitric acid on starch.

XYLONITE.—See Zylonite.

CHAPTER VIII.

PIGMENTS AND PROCESSES USED IN COLORING INDIA-RUBBER.

Most of the India-rubber goods manufactured to-day are black, this color, if it may be so called, being produced in a measure by the color of the rubber, together with the leads and other ingredients, most of which darken during vulcanization. The next prominent color, from a rubber standpoint, is white, produced by either an oxide or sulphide of zinc. Next to this range the yellows and reds, produced by sulphide of antimony and vermilion.

So many colors are unstable when brought in contact with sulphur during the heat of vulcanization, and it is so difficult to get good effects, that it is not to be expected that beautiful colors in India-rubber will ever become common. There are various methods used for changing the natural color of India-rubber. The usual way is by incorporating, by mechanical mixture, earthy pigments or metallic oxides or sulphides, or vegetable coloring matters, which, by their covering property and strength, give to the India-rubber their own particular shade. There are other methods, however. For example, there have been produced anilines soluble in benzine, that are used for surface work, such coloring being really an elastic enamel. Toys and minor articles that are ornamented in very bright colors, however, are generally painted over after vulcanization, but paint is not durable, nor does it long remain beautiful.

While it is claimed ordinarily that it is impossible to dye India-rubber, it should be remembered that the attractive colors that appear on childrens' toy balloons and similar pure gum goods are applied as dyes, the colors being analines, with methylic alcohol as a base. These colors are boiled in rainwater, and when the solution is cold the balloons are put into the coloring liquid and turned so as to have their entire surface wetted. After that, they are dropped into cold water, which washes off the superfluous color. When this is done properly, the rubber does not give off any stain at all after the first washing. The colors used in this way are red, green, blue, orange, and pink, but other shades are equally available.

In Germany a full line of aniline colors soluble in benzine is now manufactured, and for surface coloring of rubber goods they have been found very valuable. Although they are not absolutely fast, they are sufficiently so for all practical purposes. In many cases, these aniline colors, being soluble in benzine, can be mixed right with the India-rubber—that is, when it is used in the form of solution. If the product is cured in open steam heat with sulphur, some very curious effects are likely to be obtained. This was proved some years ago when a line of rubber colors was put on the market in the United States, with white oxide of antimony as a base, and anilines to give various shades. It does not often happen, however, that a problem of this kind confronts the users of aniline colors in rubber, the more general and sensible way being that of surface coloring. This is done in some cases by simply brushing the aniline color dissolved in benzine over the surface of the article. It is desirable, however, first to dip the goods in the dissolved mordant, and then to use the brush, if necessary. Where a high polish, or a polished effect is desired, some sort of elastic lacquer must be put on over the coloring matter. A very thin India-rubber solution is often used for this.

In speaking of anilines, it must be remembered that those that have to be worked up with acids should be avoided for rubber work, but there are so many others that there is no need of the rubberman making this mistake. Where colors are to be printed upon rubber surfaces, a little dextrine is added to the aniline dissolved in benzine, and to make the color dry faster, a little sulphate of manganese mixed with half of 1 per cent. of alum and added to the mass is advisable.

Black, blue, red, yellow, and green anilines are also used in coloring rubber cements that go to the leather shoe trade. These and other anilines are also used very generally in artificial leather compounds. Aniline, black, is used in water varnishes for luster coats and blankets.

It is also a good idea to sponge the rubber surface with a water solution of alum before the color is applied. The use of alum as a mordant may be supplanted by bisulphate of soda, if it is desired. The best colors available in the aniline series are reds, particularly magenta reds, and the marine and alkali blues.

A great many methods of surface coloring have been devised, some of them being ludicrous attempts at dyeing rubber. The surface of rubber is, of course, not easily affected by colors, unless it has first been attacked and roughened by some powerful solvent. Malcolm's process for this surface coloring is perhaps as harmless as any. This method is to expose the rubber to the sunlight while it is immersed in alcohol. When the surface is somewhat disintegrated, the rubber is taken out, washed, and dipped in a dye solution.

The colors that follow are described very briefly, and most of them are such that any rubber manufacturer can easily secure them for use or for experiment.

WHITE.

ONLY a few colors are available for use in making white rubber goods. Of these, the zincs take the lead, being by far the most constant and valuable. They lend their color to the mass simply by their presence as dry paints with strong coloring qualities.

OXIDE OF ZINC is used more than any other coloring matter in the production of white rubber. It is especially valuable because during the process of vulcanization it increases the whiteness of the goods. This is because the part of the zinc oxide that is turned into the zinc sulphide is a stronger white than the first. Oxide of zinc made of pure spelter is the best. Where lead and zinc ores are found together it sometimes happens that the oxide contains a certain amount of lead, and then its value as a coloring matter is injured. It is prepared by two processes, an air blast, and a steam current; in other words, by a dry and a wet process. That prepared by the wet process, even when strongly heated, contains more water than does that produced by the dry process. The specific gravity of zinc oxide is 5.61. A certain percentage of this oxide is often added to dark colored goods to increase the resiliency of the rubber. It also increases the hardness of a compound where soft gums are used. Manufacturers of insulated wire find that it increases the insulating qualities of rubber when added in moderate quantity.

ZINC WHITE.—See Oxide of Zinc.

Sulphide of Zinc.—This is a white that is fully equal to the popular oxide, and does not alter its tint under the influence of sulphur and heat. It is said to exert a distinctly preservative action upon India-rubber. Sulphide of zinc, pure and in combination with other materials, and under various names, has been sold very largely to rubber manufacturers. It is deemed especially valuable in white goods cured with dry heat. It is used in high grade white stocks, and even in pink dental rubber. It also assists in the vulcanization of rubber.

Oleum White.—A high grade of sulphide of zinc, in which is a certain proportion of blanc fixe. It is a trifle heavier than a pure sulphide of zinc, but in practice has been found to be equal if not better than either the sulphide or oxide of zinc in the manufacture of certain white rubbers.

Carbonate of Zinc.—This is a form of zinc rarely known today in rubber mills. The first white rubber, however, was made of it under a patent granted to that eminent rubber manufacturer, the late Henry G. Tyer. It is a white powder, and is a mixture of equal quantities of sulphide of zinc and carbonate of sodium, and subsequently the boiling of the same for a short time.

Borate of Zinc.—A zinc salt, precipitated by 20 to 30 per cent. of a soluble borate, the result being a white powder, which is claimed to have a distinctively preservative influence when used in rubber, while the tensile strength of the gum is much enhanced.

[Lascelles-Scott.]

Calamine White.—This is prepared from the native carbonate of zinc, by calcining and grinding. It is not a strong white, and is not nearly as good as the oxide or carbonate of zinc as a coloring matter. For a cheap white, and a filler, however, it is useful. Although the German anti-poison act of 1887 prohibits the use of zinc as a coloring matter, it does not apply to its ordinary use in rubber compounding. They rule that zinc compounds not soluble in water may be used in rubber when and where the coloring matter is mixed in the mass before vulcanizing, or as a color layer on the surface if it is covered with a lacquer varnish.

Barium White.—This is also called constant white, and comes from the sulphate of barium or heavy spar. In treatment, it is ground very fine, treated with hot hydrochloric acid, washed, dried, sifted, and then forms a fairly white, dense, impalpable powder. The pure article, obtained by precipitation, is a brilliant white, and is often used in rubber compounding. It is one of the few metallic colors that the German anti-poison act allows manufacturers to use in any way they please.

GRIFFITHS'S WHITE is a sulphide of zinc of English manufacture, prepared by precipitation, and containing a certain proportion of magnesia.

FARD'S SPANISH WHITE.—Also known as Pearl White. A tri-nitrate of bismuth, and a white that, it is said, has a future in rubber compounding. It is not easily affected by atmospheric influences, or by the action of sulphurous compounds.

[A. Camille.]

LITHOPHONE.—A sulphide of zinc in which is found a certain percentage of barium. It is a constant white, and is largely used instead of oxide of zinc for white goods, particularly in the manufacture of druggists' and surgical sundries.

BLACK.

THERE are more methods of getting black rubbers, than almost any other color, as the tendency of the gum itself is to darken under heat and the action of sulphur, and the sulphides of most materials that are used in the compounding have the same effect. Most rubber goods are made up without regard to color, and are usually a dirty brownish-black, tempered by the yellow of the sulphur bloom. Where a genuine black is wanted, however, some of the vegetable blacks or perhaps certain of the leads are employed. Lampblack is one of the most common ingredients used.

LAMPBLACK.—Pure Lampblack is pure carbon, as indeed is the diamond. Lampblack, however, is carbon in its amorphous or spongy form, while the diamond is crystaline. It is obtained on a large scale by collecting the smoke produced during the combustion of oils, fats, resins, coal, gas, tar, wood tar, petroleum residues, dead oil, and even bituminous coal. This accounts for the various grades that are to be found on the market. Large quantities of Lampblack have also been manufactured from natural gas. There are many types of Lampblack, the best in the world being employed in the preparation of Indian ink. This is made from burning camphor, a lower grade being made from the mixture of camphor and other oils. The smoke is collected on leaves, washed, dried, and sifted with the utmost care. The lines of rubber goods in which it is generally found are rubber boots and shoes, surface clothing, and carriage cloth, druggists' sundries (where the leads are deemed dangerous), and in certain compositions where emery is the chief ingredient used for grinding or polishing. A curious fact about Lampblack is that a little bit of it in unvulcanized, erasive rubber, seems to assist the erasive quality, and does not cause smutting. A little of it is also sometimes added to churning mixtures that do not readily mix. The following analysis of the composition of lampblack is given by Braconnot:

Carbon	79.1
Water	8.0
Resinous matter	5.3
Bituminous matter or pitch	1.7
Sulphate of ammonium	3.3
Sulphate of calcium	.8
Sulphate of potassium	.4
Chloride of potassium	traces.
Phosphates of calcium and iron	.3
Siliceous or earthy matter	1.1
Total	100.0

The analysis of lampblack from a large black manufactory in the United States:

Carbon		79.1
Empyreumatic resin	soluble in alcohol	5.3
	insoluble in alcohol	1.7
Humin		0.5
Sulphate of ammonium		3.3
Sulphate of lime		0.8
Sulphate of potash		0.4
Phosphate of lime		0.3
Water		8.0
Chloride of potassium		trace only.
Sand (accidental)		0.6
Total		100.0

BONEBLACK, also called animal charcoal and sometimes ivory black, is a black powder obtained by grinding the product of bones that are burned at a red heat in close vessels. It resembles vegetable charcoal, but is more dense and less combustible. A good quality should have an even color, of a rather dull shade. On analysis, boneblack shows the following:

Phosphate of lime	78.0
Phosphate of magnesia	1.5
Carbonate of lime	8.5
Carbon	10.0
Impurities, silica, iron, etc	2.0
Total	100.0

Sulphide of Lead.—This is a valuable coloring matter for rubber, as it gives a good black, besides which it makes goods exceedingly resilient. There are great differences in the production of lead sulphides, but, as before remarked, a good one is of special value to rubber manufacturers. (See Leads.)

Mineral Black is a pigment that is said to be made from bituminous lignite. It is very porous, and is not recommended for rubber work. A very little ultramarine blue added to a black in rubber, sometimes overcomes the grayish shade.

Sulphide of Uranium.—A fine black pigment more intense than plumbic blacks. It is a permanent color, and is said to be a preservative of rubber.

Black Hypo.— This is also known as hyposulphite of lead. It is really a mixture of thiosulphate of sodium mixed with acetate of lead, and appears as a fine white crystaline precipitate, which should be called thiosulphate of lead. There are two forms, the white hypo and the Black Hypo, the difference being that the white when heated is transformed into a soft black powder containing very little free sulphur. The black of the compound being sulphide of lead often contains over 90 per cent. of pure sulphide. It is an excellent vulcanizing agent, and also a filler. When properly prepared it makes goods absolutely free from bloom.

Carbon Blacks of late have been used very largely in rubber compounding and have done excellent work. They are not as black, as a rule, as the better grades of lampblack made from oils or resin. They are in many cases wholly inert, however, and therefore perfectly safe to use. One of the best types of this sort of coloring matter comes from a graphite mine in the United States. It is wholly amorphous, and has none of the flaky make-up that ordinary graphite has, and is 97 per cent. pure carbon. Carbon Blacks, it is also said, give a brighter finish to varnished goods than ordinary lampblacks.

OAK BLACK.—A product of the distillation of oak wood after draining off (1) wood alcohol and (2) a product resembling tar. It is used in certain black insulating compounds in connection with shellac, coal tar, paraffine, and asbestos.

BLUE.

BLUES are not largely used in general rubber work. They are found chiefly in toys, in sheetings, and in certain packings. The most important blue is—

ULTRAMARINE.—This is made from lapis lazuli. The exact composition of this coloring matter is not known, but it is said to be based on a silicate of alumina with sulphide of sodium. An artificial ultramarine is often produced which is equal and often superior to the natural pigment. This is made of kaolin, carbonate of sodium, willow charcoal, and sulphur. The following analysis of natural Ultramarine is given:

Silica	37.6
Alumina	27.4
Sulphur	14.2
Soda	20.0

Analyses of the best artificial Ultramarines show these figures:

Silica	40.25	39.39	40.19
Alumina	26.62	24.40	25.85
Sulphur	13.42	12.69	13.27
Soda	19.89	21.52	20.69

Ultramarine appears in commerce as a fine blue powder of various standards of fineness. Acids readily destroy it, but alkalies have no effect on it. It stands heat well, not changing below a low red. It is used in cements for backs of memorandum blocks, and in blue soft rubber goods, particularly in vapor cured goods, such as sheeting. When mixed with chrome yellow it makes a green; with colcothar, it makes a violet. Mixed with rose pink, oxide of zinc, and Indian red, it produced the well-known wine-colored coat that was so popular a few years ago. It is claimed that Ultramarine blue keeps rubber from overcuring, and that it is, therefore, a most useful ingredient to add to compounds that are exposed to heat.

YALE BLUE.—In certain soft rubber goods, where a strong blue is needed, ultramarine was found unsatisfactory. A firm of rubber chemists therefore produced Yale Blue, which is a strong coloring matter, and wholly inert as far as the rubber is concerned.

SMALTS.—This is what may be called a deep tinted cobalt glass. The analysis of Smalts of good quality is as follows:

	Deep-colored Norwegian.	Pale-colored German.
Silica	70.9	72.1
Potassa (with traces of soda and lime)	20.4	20.0
Oxide of cobalt	6.5	2.0
Alumina	.4	1.8
Peroxide of iron	.3	1.4
Other earths and oxides, and loss	1.5	2.7
Total	100.0	100.0

This is one of the few colors that are practically indestructable. In using Smalts for the pigment, large quantities are necessary, as the color is not exceedingly strong.

COBALT BLUE is manufactured from oxide of cobalt, phosphate of cobalt, and alumina. It is rarely used in coloring rubber where the ingredients are to be mixed with the mass, ultramarine being much superior. Also called Smalts.

Thenards blue is similar to cobalt blue, but is a more beautiful pigment. It is used chiefly as a surface color. White pigments in small quantities added to this blue make beautiful turquois colors.

PRUSSIAN BLUE.—A dark brilliant blue compound, having iron for a base. There is a soluble and an insoluble variety of this compound which is of a somewhat complex chemical constitution. Heated strongly in the air, the insoluble form of Prussian Blue burns like tinder. When boiled with caustic potash, it is decomposed. If the dry powder be strongly rubbed in a mortar, it assumes a copper red luster. In commerce it occurs in irregular shaped masses, having a characteristic conchoidal fracture and copper red luster.

CHROME BLUE is manufactured from silica, fluor spar, and chromate of potash. The resultant material is a deep blue, vitrious mass which is reduced to an impalpable powder. It is less sensitive to acids than ultramarine, and is better adapted for rubber goods. [Jules Garnier.]

MOLYBDENUM BLUE.—A pigment recommended by Lascelles-Scott, which is a natural bisulphide of molybdenum, found chiefly in Sweden. It is an exceedingly beautiful blue, but at present is rare. The distinguished chemist above quoted mentions that large new deposits of this mineral have been found in the United States and Australia, and that it is likely to be so cheapened that it will be a valuable rubber pigment.

INDIGO BLUE is prepared from plants of the *indigofera* genus. Pure Indigo is insoluble in water, nor is it soluble in weak acids or alkalies. A small percentage is dissolved in alcohol and its solution is more considerable in turpentine. Indigo Blue for rubber is said to be valuable on account of its preserving qualities, which are double that of other blues.

RED AND BROWN.

THE strong red coloring matters used in rubber work are mostly of a mercurial base. These are vermilion, red chromate of mercury, sulphide of mercury, and iodide of mercury. The Chinese vermilion, which is the best, is prepared by a special process of their own, and contains 89 per cent. of pure mercury, the rest being sulphur. This coloring matter is used very largely in dental vulcanite, small amounts of it also giving excellent shades in soft rubber goods. Cinnabar and Paris red are also mercurial sulphides, and very strong colors. The sulphides of mercury are really the only ones that are safe and valuable for producing these colors. Red chalk and natural clay containing a certain amount of iron are used chiefly as fillers in rubber goods, although a certain quantity of them produce a dark red color.

VERMILION.—The red form of mercuric sulphide is a scarlet red powder of specific gravity 8.124. It is sometimes adulterated with red lead or red oxide of iron, but such adulterations can be detected by heating a small sample of the suspected article on a porcelain or platinum dish. If any adulterant is present it will remain behind as a residue, since pure Vermilion is completely volatile. This substance is sometimes called cinnabar. A substitute for vermilion in hard rubber was brought out by John Haliday in 1870. This was a mixture of garancine and cochineal, in water solutions, boiled and mixed in the proportion of 5 parts of garancine liquor to 1 part of cochineal liquor. To each gallon of this compound liquor 2 pounds of pure oxide of antimony was added; then, after heating until the water was evaporated, the new coloring matter perfectly dry. Another substitute for vermilion was white oxide of antimony. According to A. D. Schlesinger, the veteran of hard rubber experts, white oxide of antimony, when mixed with India-rubber and sulphur, will, during vulcanization, impart to hard rubber a light red color very similar to that obtained by the use of vermilion. The proportion of sulphur is the same as is used ordinarily in making vulcanite, while to each pound of rubber is added 12 ounces of antimony sulphide.

RED OXIDE OF IRON.—This is familiar as iron rust. It is artificially prepared and forms a scarlet powder of a specific gravity of 4.46. This contains about 5 per cent. water of crystalization, which cannot be driven off at temperatures up to 212° F., and with difficulty at higher ones. (See Colcothar.)

PEROXIDE OF IRON.—An old name for the sesquioxide of iron, now called ferric oxide. (See Oxide of Iron.)

PRINCE'S METALLIC PAINT.—An oxide of iron.

INDIAN RED.—Another name for oxide of iron.

RED HEMATITE.—An ore of iron, somewhat soft and friable. Specific gravity 5.19 to 5.28. Composition 70 per cent. iron, 30 per cent. oxygen. Insoluble in water, alcohol, or rubber solvents. As a colorant in rubber work it is unchangeable chemically. Used in packings and for dark maroons.

VENETIAN RED.—See Colcothar.

RED OCHRE.—An impure oxide of iron. A dull red earthy substance containing clayey matter, and having a specific gravity of about 5.2. Used chiefly as a filler, as the color is not strong. As far back as the time of Dr. Mattson, Red Ochre, venetian red, and Indian red, were advised by him for use in rubber compounding. Indeed, he obtained a patent for packing in which venetian red was the principal adulterant.

ORANGE VERMILION gives a very handsome color in connection with rubber, but is rarely used, as it is not permanent if other metals, such as copper, brass, iron, and zinc, come in contact with it.

CRIMSON SULPHIDE OF ANTIMONY.—This is altogether the best antimony color now in use. It not only gives a fine shade of orange or red, but it also is an excellent vulcanizing agent.

COLCOTHAR.—A form of oxide of iron of the specific gravity of 4.8 to 5.3. It is the residue left in the manufacture of fuming sulphuric acid from green vitriol. The least calcined portions, which are scarlet in color, are termed jewelers' rouge, and the more calcined parts, of a bluish shade, are called crocus. Its composition is that of ferric oxide. In its reaction it is indifferent, being very stable under ordinary conditions. Colcothar is a dull red and is often used in red packings, soleings, etc. Many rubber chemists prepare their own Colcothar, as they are able to get brighter shades than is possible from the goods ordinarily sold in the open market.

UMBER.—A brown earthy mineral, containing chiefly the oxides of iron and manganese. The following analysis, by Prof. A. H. Church, is taken from a choice specimen of Cyprus Umber: Oxide of iron, 48; oxide of manganese, 19; silica, 13.7; water yielded at a heat of 212° F., 4.8; mixture of lime, magnesia, alumina with organic matter, 14.5. In using Umber for rubber compounding, care should be taken to dry the material thoroughly at 212° F., before it is used. Burnt Umber is the product obtained by roasting the above material. It is slightly redder in color and will naturally contain less water. For brown colors, in addition to Umber, various natural earthy matters are used, as are also oxy-sulphide of antimony and sepia, the latter being an animal coloring matter made from the bright fluid formed in the ink bag of cuttle fishes. Sienna and chestnut brown are practically the same as Umber, while Vandyke brown is made of oxide of iron, ground very fine, and is not injurious to rubber. While these ingredients are practically inert, they do not make the best of rubber compounds, as the resulting compound is apt to have a hard stony feeling.

YELLOW.

YELLOWS are not often demanded in rubber work, except in a few fancy articles and in hose markings. The most common is that produced by the golden sulphuret of antimony, but color is not what is sought in the use of that ingredient, but rather the excellent rubber produced by it when used instead of sulphur. Other mineral yellows used are strontium, chromium, cadmium, barium, and arsenic. Chrome yellow is made from a lead base which darkens when subjected to vulcanization.

CADMIUM YELLOW.—This is the best pigment for producing yellow in a rubber compound. It does not injure the elasticity or strength of the India-rubber in any way, and, while it has no special effect on vulcanization, perhaps hurries it a little. It is not injurious to the health of persons using it, and is generally used for surface ornamentation of toys, etc. It is sometimes mixed with yellow sulphide of tin to cheapen it. While Cadmium was ruled against in the German anti-poison act, the sulphides of this metal were made an exception, and said to be safe. In dental plates, however, where the coloring matter was used in large quantity, it was advised against. The costliness of Cadmium Yellow at present bars its general use in rubber.

AUREOLIN YELLOW.—A very handsome color, and one that is stable and brilliant. It is made up of acetate of cobalt and nitrate of potassium. The color stands the light well, and sulphur compounds have little influence upon it. This is chiefly used for surface work.

GAMBOGE YELLOW.—Obtained from the *Garicinia morella*. It contains from 20 to 25 per cent. of gum, 65 per cent. of resin, 3 per cent. of volatile oil. It is soluble particularly in spirits, in a number of oily liquids, and partially in water. Finely pulverized Gamboge may be mixed with rubber, and is said to be a preservative of it.

BARBERRY YELLOW.—Made from the root or bark of the *Barberis vulgaris*. It is largely used in coloring leather surfaces, and, in connection with gamboge, is said to be useful in rubber work.

YELLOW OCHRE.—There are several ochres, all of them being practically oxides or iron. They are earthy substances of no particular reaction, very stable, having a specific gravity about 5. Their low cost renders them available for almost any work, but the colors produced are not especially beautiful.

ARSENIC YELLOW.—Also known as king's yellow, and is a term applied to sulphide of arsenic. A cheap grade of this, which is really only an imitation, is manufactured by mixing together litharge and white arsenic, and grinding the product. Either of these, of course, is poisonous, and they are very rarely used or needed in connection with rubber. The specific gravity of Arsenic Yellow is 3.48. Although a sulphide, there is not enough sulphur in its composition to vulcanize India-rubber. On account of its poisonous properties, this yellow has been largely superseded commercially by the comparatively harmless chrome yellows. Another name for this color is orpiment. It was often used in rubber compounds of twenty years ago. A small quantity in white zinc stock takes off the glaring white effect, and produces a handsome cream white. Must be in an impalpable powder to bring out the color.

CHROME YELLOW.—Ordinarily the chromate of lead, which is largely used as a pigment. It is somewhat poisonous and is apt to oxydize organic substances, particularly if sulphur is present. Has been used in the surface ornamentation of rubber toys, but such use is generally condemned. The only Chrome Yellows that are really valuable for rubber work are the chromate of zinc, or possibly the chromate of strontium.

ORPIMENT.—See Arsenic Yellow.

GREEN.

IT is fortunate that greens are not largely sought in the rubber industry, for they are rare. Arsenic greens in many cases are not to be thought of; therefore about the only ones that are available, unless very high cost goods can be utilized, are the following:

CHROME GREEN.—A coloring matter that is not affected by strong acids, or alkalies, and which is inert when mixed with India-rubber. It is the best mineral green that can be used in connection with rubber. It is really a sesquioxide of chromium; and may be mixed with rubber, with any kind of solvent, and with other oxides and pigments, without hurt to the compounds.

TERRA-VERTE is of mineral origin, and is imported in large quantities from Italy. It is a pale neutral green of moderate cost, and is not injurious to rubber. On analysis it shows:

	No. 1.	No. 2.
Silica	51.50	46.00
Alumina	12.00	11.70
Protoxide of iron	17.00	17.40
Lime	2.50	3.00

Magnesia	3.50	8.00
Soda	4.50	....
Water	9.00	13.90
Total	100.00	100.00

The analysts of the above were, of No. 1, Klaproth; of No. 2, Berthier.

GREEN ULTRAMARINE is made by a process very similar to that made in producing blue of that name, and its action upon rubber is almost identical with that of ultramarine blues.

CHAPTER IX.

ACIDS, ALKALIES, AND THEIR DERIVATIVES, USED IN THE RUBBER MANUFACTURE.

As a rule neither acids nor alkalies, in the strict sense of the term, are largely used in ordinary rubber compounding. In a great many of the processes, however, that go far to make up finished goods, acids are used, as, for example, in those employed in the reclaiming of rubber chemically. Alkalies also are most necessary, a notable example being the use of caustic potash and caustic soda solutions in removing sulphur from manufactured goods. A great variety of uses other than these are indicated in the following pages:

ACETIC ACID.—This is usually obtained by the dry distillation of wood fiber, peat, or sawdust. The strongest form is known as glacial and occurs in large watery crystals, readily liquified. The common commercial acid usually has a brown or yellowish color, due to impurity, since the pure acid is colorless. Its specific gravity is 1.05, and it has a characteristic odor familiar enough in vinegar. As an acid it is not very corrosive, and its compounds are easily decomposed by mineral acids. It is quite volatile. The primary use of this acid in connection with India-rubber is in the coagulation of rubber milk. It is a prominent component part of the smoke used in coagulating fine Para rubber. It has also been used under the Vaughn process for coagulating Balata, and in the manufacture of certain substitutes like linoxin, Parkesine, etc.; in connection with nitro-cellulose and castor oil in the production of certain waterproofing compositions; by Brooman in separating whiting, white lead oxides, etc., from vulcanized rubber; and in shoemakers' blackings in connection with caoutchouc oil, vinegar, molasses, and lampblack.

ALE.—A beer made from malt, distinguished chiefly by its strength and the quantity of sugar remaining undecomposed, which enables the liquor to keep, without requiring a large amount of hops. A mixture of ale and linseed oil, in the proportions of 8 parts ale to 2 parts linseed oil, is used in dissolving isinglass, in which is afterward incorporated shellac and India-rubber in the formation of what is known as ale cement.

ALUM.—A general term for several chemical compounds of aluminum, potassium, chromium, and ammonium. Common alum is the double sulphate of potassium and aluminum, having a specific gravity of 1.7 and containing 45 per cent. of water of crystalization, one-quarter of which is expelled on heating to 140° F. It is soluble in water 9½ parts per 100 when cold, 357 parts per 100 when hot. Chrome Alum is a double sulphate of chromium and potassium, its specific gravity being 2.7, and containing 43 per cent. water of crystalization, which is almost entirely lost at 392° F. It occurs as dull purple crystals, slowly soluble in water to 20 per cent. in the cold and 50 per cent. in hot water. Its action on gelatine is remarkable for its hardening qualities. Ammonia Alum, the double sulphate of aluminum and ammonia, is largely used in place of common alum. It contains 48 per cent. of water of crystalization and has a specific gravity of 1.63. Strongly heated, it yields sulphate of ammonia water and a very small quantity of of sulphuric acid, while alumina is left behind. It is soluble in water 13 per cent. cold, 422 per cent. hot. Roman Alum has the same general characteristics as common alum, but contains a little more alumina.

Alum is used in many of the shower-proof mixtures for cloths of the cravenette order, that are to-day bought and made up by manufacturers of mackintoshes. It is also sometimes used in the manufacture of sponge rubber. By Garnier's process it is also used in spirituous solution to cure rubber without heat by mixing with it. Used also in Wray's substitute for Gutta-percha. Alum was used in Payne's Gutta-percha compounds for proofing, varnishing, and paints. Ghislin, who prepared some curious compounds from seaweed and India-rubber, mixed alum, gelatine, and metallic oxides in his compounds. It is also sometimes used in compounding rubber to make sponge effects and mixed with sulphate of iron and soap, in a water mixture with boiled linseed oil, to make flexible waterproofing compounds.

AMMONIA, at ordinary temperature, is a colorless gas of well known odor and sharp biting taste. It is usually met with in the arts in watery solution, the specific gravity of which varies with the amount of ammonia gas dissolved. The strongest, sometimes called caustic ammonia, contains 32.5 per cent. of the gas, and has a specific gravity of .875. Ordinary commercial ammonia has a percentage of 9.5 and a specific gravity of 0.96. The weakest usually has a percentage of 5.5 and a specific gravity of .978. Ammonia has a powerful solvent action upon sulphur, is alkaline in its nature, and very volatile, so that much care is requisite in handling it. It has long been known to have a preservative effect upon India-rubber; for example, low grade African rubbers are often treated with Ammonia to neutralize the smell, and also to toughen the rubber. In the cold-curing process a saucer of Ammonia put in the bottom of the vapor room will effectually neutralize the fumes of chloride of sulphur. It is also advised to wash vulcanite that has begun to perish with an Ammonia solution. Soft rubber goods also are preserved, according to Dr. Pol, by the immersion for an hour in a solution made of 1 part of ammonia, and 2 parts of water.

Sievier dissolved India-rubber in Ammonia, leaving it in a closed vessel for a long time, after which he heated the solution and distilled the Ammonia gas in cold water. Concentrated liquor of Ammonia is added to milk of the rubber tree to preserve it for transportation. Where vegetable fibers are reduced to cellulose and mixed with India-rubber, the rubber is first steeped in Ammonia and then dissolved in some suitable solvent. Newton mixed Ammonia with India-rubber and Gutta-percha, and then treated the gum with chlorine, making a white hard compound which he claimed would stand all varieties of climates, acids, greases, etc.

Aniline.—A colorless oily liquid, manufactured chiefly from coal tar or nitrobenzene. It is a base from which the brilliant aniline dyes are made. Aniline used by Parkes in the manufacture of Parkesine, is also a solvent for Gutta-percha.

Arsenate of Potash.—It is a very soluble compound of arsenic with potash and forms what is known as Fowler's solution. In the dry state it is a white powder soluble in alcohol up to 4 per cent. Arsenate of Potash was used by Forster, among his earliest experiments, to partially vulcanize a compound made up of India-rubber and shellac.

Barium Chloride.—A white crystaline powder, insoluble in alcohol but soluble in hot water, 78 per cent., and in cold 38 per cent. Its specific gravity is 3.05. It is not of great technical importance, its principal value being that of a test for sulphuric acid. To makers and users of sulphurets it affords a ready means of determining the presence of free sulphuric acid, so liable to occur in these bodies and so injurious to rubber compounds when present. A suspected sulphuret should be boiled for a moment with a little distilled water, the water filtered off, and a drop or two of a solution of Barium Chloride added; a white cloudiness that will settle in the form of a white powder proves the presence of sulphuric acid and such a sample should be rejected. Barium Chloride is a powerful poison. Used with size and acid resin as a shower-proof mixture.

BISULPHATE OF POTASH.—A white powder obtained as a by product in chemical manufacturing. Soluble in twice its weight of cold water, and in half its weight of boiling water. It contains sulphuric acid so loosely held in combination that it is driven off upon heating. Its specific gravity is 2.16. (See Potash.)

BICHROMATE OF POTASH.—The principal compound of chromium, which occurs in the form of orange red crystals, that are soluble in water and are largely used in dyeing. Mixed with sulphuric acid, it is used in bleaching palm oil and other fats. Bichromate of Potash is used in vulcanizing the compound known as elastic glue; also used in Christia gums.

BLEACHING POWDER.—See Chloride of Lime.

BORACIC ACID.—This is found native in the vapor which arises from certain volcanic rocks in a saline incrustation in volcanic craters and in combination with borax. It appears in the form of pure white leathery crystals. Boracic Acid is used with tungstate of ammonia, Kauri, borax, and India-rubber in the production of the woodite fireproof compositions.

BORAX, or BIBORATE OF SODA.—Sometimes also called tincal; a compound of soda and boracic acid. The purified commercial article contains about 47 per cent. of water of crystalization and is usually in the form of large odorless crystals, or a white powder obtained by grinding. The crystaline form has a specific gravity of 1.69. Borax is quite soluble in water, but not in alcohol or any of the common solvents for rubber. At a moderate heat Borax loses water, and separates as a spongy mass called calcined borax, while at a higher heat it melts into what is known as borax glass. Immense deposits of it are found in the United States, and it is also found in India, Hungary, and other parts of the world. A good waterproof cement is made of a mixture of Borax and shellac boiled in water. Borax, or a solution of biborate of sodium, has the property of dissolving many resins. Lascelles-Scott describes the manner in which an emulsion of rubber may be preserved by a Borax solution. To a solution of rubber, in any one of the common solvents, a small portion of alcohol is added. This is mixed with a 2-5th saturated solution of Borax, previously heated from 120° to 140° F. This is agitated until the temperature has cooled down to the temperature of the air. From 3½ per cent. to 4½ per cent. of India-rubber should be present in the fluid when finished. A higher strength quickly separates and sometimes causes the entire quantity to coagulate. Madagascar or Sierra Leone rubbers are advised for Borax solutions. Solutions of borated rubber are adapted for waterproofing and for preserving mats, marine bedding, etc. Borax is also advised for preserving rubber milk from coagulation. It is also an important ingredient in the water varnishes used for luster finish, for surface coats, army blankets, etc.; is used in waterproofing compounds composed of rubber, boracic acid, Kauri, tungstate of ammonia; mixed with Gutta-percha and shellac, it was used by Hancock as an insulating material.

Carbolic Acid, also known as Phenic Acid, is obtained chiefly during the destructive distillation of coal. The liquid has a hot burning taste, and is largely used for its antiseptic qualities. If white crystalized carbolic acid is added to the paste from which matrices in rubber stamp making are manufactured, it preserves the mixture for a long time. Carbolic Acid is used as a preservative of rubber sap, where it is coagulated by the process employed by The Orinoco Co., in Venezuela. Carbolic Acid has also been used in connection with a little ammonia to increase the elasticity of low grade African gums, being used as a solution before the gums are washed. It is also used for treating fabrics, such as hose linings for fire and mill hose, to prevent deterioration and rotting. Used in certain fiber-made substitutes.

Carbonate of Ammonia, obtained during the dry distillation of bones, is a white crystaline powder of very penetrating smell, from which quality it takes its popular name of smelling salts. Exposed to the air, it yields ammonia and absorbs water, becoming superficially converted into bicarbonate. It is used industrially for the removal of grease from cloth and cleaning woolen fabrics. Carbonate of Ammonia is used also in the manufacture of sponge rubber, and in hollow work, where its expansive force is utilized to effectually mold the article.

Carbonate of Soda.—Also called Sal-soda, washing soda. Prepared from cryolite, salt, etc. Its specific gravity is 1.45, when crystalized. The crystaline form contains 64 per cent. of water of crystalization, of which one-half is driven off by gentle heating. It is a white crystaline substance and alkaline taste. It is found in the ashes of many plants, is produced artificially in large quantities from common salt, and is used as an alkaline agent in many chemical industries. Rubber, burnt umber, Japan, and a coloring matter are mixed with a certain proportion of Sal-soda for a waterproofing composition. Under the common name saleratus, Carbonate of Soda is used as follows: Instead of sunning surface goods, like rubber coats and blankets, they are often brushed over with a mixture of saleratus and powdered charcoal right after the stock leaves the calender. Sometimes the saleratus is left out, and only the charcoal is used.

Caustic Soda.—The chief use of this, in the manufacture of rubber goods, is in the dissolving of sulphur that is formed on the surface of goods, and which is known as bloom. According to H. L. Terry, F. I. C., the bulk of the alkali supplied to rubber manufacturers in England is used in removing the sulphur from elastic thread. Of course it is used in treating tobacco pouches, fine sheet articles, and blacks, reds, or maroons, that should have a good clear color. The boiling of rubber goods is usually done in wooden tanks in which steam can be passed, and sometimes in slate tanks, as iron is attacked by the alkali. On good grades of rubber caustic soda has no action at all; where a large quantity of resin is present, however, it may dissolve some of them, forming resinates of soda. Heavily compounded rubbers, whether they contain substitutes, gums, or compounds, unless they are absolutely inert, are also liable to be attacked through the dissolution of their ingredients. Camille describes a process whereby shoddy is treated with a solution of carbonate of soda in devulcanization. In this, the rubber is boiled several hours in a solution of caustic soda, the result being that it will sheet when the process is completed. Rostaing purified Gutta-percha by boiling several hours in caustic soda, or in a mixture of caustic soda and potash in water.

CATECHU, or CUTCH.—Known formerly as Japan earth. Made from the sap of an East Indian tree, and used chiefly in dyeing. Is very astringent, and is soluble in water. It appears in commerce in dark brown irregular lumps. Contains 40 to 50 per cent. of a peculiar tannic acid. Used in packings and goods made from the whaleite formulas. Johnson's artificial leather was made of catechu, rosin oil, linseed oil, turpentine, and starch, mixed with a little hot Gutta-percha. A number of other compounds, both with and without India-rubber, contain catechu, but chiefly those which were compounded from gelatine, starch, and gluten. Catechu is mixed with Gutta-percha in solution in order to make it harder.

CAUSTIC AMMONIA.—See Ammonia.

CAUSTIC POTASH,—As occurring in commerce, it is a white solid substance of the specific gravity about 2.5. It is hard and brittle, and very destructive to animal or vegetable substances. It rapidly takes up water from the air, and may be used to obtain a dry atmosphere in a confined vessel. It is also a greedy absorbent or carbonic acid, becoming converted into the carbonate thereby. Solutions of potash should be clarified by allowing impurities to subside. Its taste is bitter and acid and its smell unpleasant. Alcoholic Caustic Potash is used in analysis of vulcanized India-rubber and was introduced by Henrichs, particularly to separate India-rubber from India-rubber substitute. Caustic Potash is mixed with flowers of sulphur for boiling drawing rolls, the potash making the rubber more solid, while the sulphur gave a peculiar surface, making it better for drawing. Used in water solution to remove bloom from cured rubber. It is also used in certain substitutes for hard rubber, like voltit. Potash was early used in extracting the sulphur from ground vulcanized rubber. A percentage of it is used to-day in neutralizing the acid used in the chemical recovery of rubber.

CHLORIDE OF AMMONIUM.—Also known as muriate or hydrochlorate of ammonia, or sal-ammoniac. Obtained largely from gas works. Specific gravity 1.5. Usually occurs in small crystals of a sharp, saline taste. When dissolving in water a considerable reduction of temperature occurs, and this has rendered it valuable for cooling purposes. At temperatures above 212° F. it is completely evaporated, and a decomposition occurs into ammonia and muriatic acid. It is used in certain packings in which iron filings are incorporated.

CHLORIDE OF CALCIUM.—A crystaline substance containing about 50 per cent. of water of crystalization, which is lost on heating to 392° F. The specific gravity is 1.61, and that of the dried form 2.21. Its extreme attraction for water makes it useful in obtaining a dry atmosphere in any closed receptacle. Its color is white, taste acrid and sharp. It absorbs ammonia readily and will give it up again on heating. It is used in bookbinders' cements.

CHLORIDE OF LIME.—Sometimes called bleaching powder, although this latter is a mixture of the chloride and hypochlorite of lime. Industrially, its chief use is for bleaching purposes, dependent upon the amount of chlorine it contains. Commercial bleaching powder is a white powder with a faint smell of peculiar character and gradually becoming moist on exposure to the air, while it gradually decomposes and absorbs water and carbonic acid. Even in closed vessels decomposition occurs, and sometimes so suddenly and with such a rise of temperature that explosions occur. Hence it should always be used fresh and a guarantee obtained from the vendors (as is customary) of the quality of the article. Chloride of Lime is the basis of a cold curing process known as Caulbry's (which see). Gutta-percha boiled in it and then mixed with rosin and paraffine is used in insulation.

CHLORIDE OF SODIUM (or common salt) has a specific gravity of 2.3. It is a very stable compound, soluble in water at the ordinary temperature to the extent of 36 per cent., at the boiling point 39 per cent. At the freezing point water will take up 5½ per cent. of common salt. It is used, as is well known, in coagulating many of the rubber saps. Salt is viewed with considerable distrust by ordinary manipulators of rubber. Payne, however, treated Gutta-percha scraps by boiling water, salt, and oil of vitriol, to get a solution to which he added other gums and metallic oxides to get a waterproofing mixture. Cooley made artificial leather of Gutta-percha dissolved in resin oil, and added 25 per cent. or more of salt, to which he added starch or other saccharine substances. Salt, in the form of brine, is used in washing the compound known as tremenol as a last process. It is also used in shower-proofing compounds, in connection with paraffine and sulphuric acid.

CHLORIDE OF ZINC was known formerly as butter of zinc. It is formed by burning zinc in chlorine gas, or by dissolving it in hydrochloric acid, the solution being evaporated. The anhydrous form is a whitish gray mass which readily fuses, and can be sublimed at a high temperature. It deliquesces on exposure to the air, and is readily soluble in water, the solution having a bitter taste, and acting in a concentrated state as a powerful caustic. One of the best processes ever known for reducing the fiber in recovering rubber was that in which this substance was employed instead of acid. A boiling solution of Chloride of Zinc was used in deodorizing by Brockedon, who also mixed it with Gutta-percha, adding sulphur and vulcanizing the gum. Hancock also subjected Gutta-percha for a moment or two to binoxide of nitrogen, then immersing it in a boiling solution of chloride of zinc, which he claimed greatly improved its quality.

CHROMIC ACID is not readily obtained in a free state, but forms many well-known salts, such as chrome yellow, for instance. It is analogous to sulphuric acid. Vulcanized rubber immersed in it at 140° F., remained a month, and was apparently unharmed. It is also used in the manufacture of the substitute known as corkaline.

CITRIC ACID.—An organic acid that occurs in lemons, limes, and many other fruits. It is readily soluble in water, and has an intensely sour taste. Has been used in the coagulation of Balata. Vulcanized rubber immersed in it at 140° F., remained a month, and was apparently unharmed.

CREAM OF TARTAR.—A white crystaline substance with an acrid taste, a very common ingredient in baking powders. Is called also Potassium Bitartrate. Is made from purified tartar, or argol. Is used in artificial ivory made from resins in solution.

Crystals of Soda.—See Carbonate of Soda.

Cyanide of Potassium.—A white crystaline substance, very poisonous, of a sharp bitter taste. It is very easily decomposed, even on exposure to the air absorbing carbonic acid and yielding prussic acid, which gives the salt its peculiar smell of peach kernels. The vapors thus given off are very poisonous. Cyanide of Potassium was used by Brooman "to give clearness to the gum which was made from ground vulcanized rubber, which had been treated with alkalies and acids to remove sulphur and adulterants."

Fluoride of Silicon is a colorless gas. What is used in the arts is a solution in water, forming a very sour fuming liquid, acting like a strong acid. It is easily decomposed and may be used for etching glass if allowed to evaporate upon it under heat. It is prepared from flints or silica in some such form as sand or powdered glass. Used in treating meerschaum and paper pulp which, combined with certain resins, forms an artificial ivory.

Formic Acid obtains its name from the fact that it was first obtained from the red ant. It is a fuming liquid with a pungent odor, boiling at 212° F. It is now made from a mixture of starch, binoxide of manganese, sulphuric acid, and water. It has been suggested as an ideal precipitant for rubber milk. It is quite volatile, could be easily washed out, and would be found more beneficial to the rubber than many of the alkaline solutions now used.

Hydrochlorate of Ammonia.—Another name for Muriate of Ammonia or Sal-ammoniac. (See Chloride of Ammonium.)

Hypochlorite of Lime.—One of the principal constituents of bleaching powder. It does not exist alone. (See Chloride of Lime.)

Hydrosulphuret of Lime.—Lime that has been treated with hydrogen sulphide. It is an offensive smelling substance, of a dirty greenish grey appearance, and is obtained in the process of purifying coal gas. It decomposes easily, giving off sulphuretted hydrogen. It will absorb bisulphide of carbon and is soluble in alcohol. Its liability to oxidize should render it of questionable use in compounding. It was used by Hancock in vulcanizing India-rubber.

HYDROCHLORIC ACID is known usually by its trade name of muriatic acid. It is also known as chlorhydric acid, and spirits of salt. It is one of the principal mineral acids. Used in the arts in the form of a watery solution, of which the strength varies from a specific gravity 1.01 or 2° Beaume with 2.02 per cent. acid to 1.21 or 26° Beaume, with 42.85 per cent. acid. Each .01 increase of gravity corresponds to 1° Beaume and 2.02 per cent. of acid. It is corrosive to the skin and attacks nearly all metals. It has no action on caoutchouc and very little on oxidized linseed oil if the acid be dilute. With soda and its compounds generally speaking it will form common salt and with metals it forms chlorides thereof. Hydrochloric acid, during the treatment of reclaimed rubber, turns whiting into chloride of lime. As the chloride is more soluble than sulphate of lime much of it washes out during the vigorous cleansing that the rubber undergoes to remove the free acid. Hydrochloric Acid, according to tests made by William Thompson, F. R. S., did not at all injure India-rubber, although it was kept in it at a temperature of 140° F. for a month. Concentrated hydrochloric acid has but little action on Gutta-percha, and tubing made from it is therefore largely used in chemical factories for running this acid from one vessel to another. Hydrochloric Acid is used in the manufacture of turpentine rubber, and in one of the last processes in the analysis of vulcanized India-rubber. In preparing a hard rubber compound, Austin G. Day used linseed, cottonseed, castor, and coal oils; hydrochloric and nitric acids; bicarbonate of soda, muriate of tin, coal tar asphaltum, sulphur, and Gutta-percha.

IODIDE OF ANTIMONY.—A brownish red crystaline mass, which yields a cinnabar red powder. It is soluble in hot carbon bisulphide. Its specific gravity is 4.39. It was used by Parkes in vulcanizing India-rubber.

IODIDE OF ZINC.—A very unstable substance. A white granular powder, odorless and of sharp saline metallic taste. Chiefly used in medicine. It was used by Hancock to assist in the vulcanization of India-rubber.

LIQUOR OF FLINT.—See Silicate of Soda.

MIMO-TANNIC ACID.—See Catechu.

MURIATE OF AMMONIA.—See Chloride of Ammonium.

MURIATIC ACID.—See Hydrochloric Acid.

NITRATE OF LEAD.—A compound of lead and nitric acid containing 62.5 per cent. of lead. Its specific gravity is 4.58. It has an astringent metallic taste, crackles when heated, detonates when thrown on red hot charcoal, and takes fire when ground with sulphur. Its color is white and it is largely used in dyeing and for making chrome yellow (which see). It is used with gums in the production of shower-proof mixtures with sugar of lead and alum.

NUT-GALL.—An excrescence formed on the leaves of a species of oak called *Quercus infectonia*. It is used in the arts for the sake of the tannic acid it contains. There are three varieties in commerce—green, white, and black. The black and the green are the best. Those grown in warm countries are the best. Aleppo galls contain from 60 to 66 per cent. of tannic acid. There is a variety of nut-gall known as Chinese, imported from Japan, China, and Nepal. The gall is somewhat bean-shaped or is covered with a yellow gray felt. It contains from 60 to 70 per cent. of tannic acid. Nut-gall is used in certain places instead of tannin, which see.

NITRIC ACID.—Chemically an oxide of nitrogen. Technically a strongly acid liquid consisting of an aqueous solution of the pure acid. Its action on different bodies is various. Some, like sulphur, phosphorus, carbon, and many organic substances are easily oxidized. Tin and powdered antimony are rapidly converted into their oxides, while turpentine, if poured into the strong acid, is attacked with almost explosive violence with the evolution of light and heat. Straw or sawdust may become ignited if impregnated with this acid. Cotton wool is converted by it into gun cotton. Rubber immersed in Nitric Acid at a temperature of 140° F. was injured in a few hours, and in a few days its elasticity was destroyed. while at the end of the month it was reduced to a pulp. Nitric Acid attacks Gutta-percha very powerfully, and evolves suffocating fumes of a deep red color, the gum meanwhile being reduced to a pasty mass which afterwards dries and becomes very brittle. According to H. L. Terry, F. I. C., Nitric Acid of any strength has a very deleterious effect upon India-rubber, the action of the fuming acid being to form immediately an oxidized body of a resinous nature. He holds, therefore, that the weaker acid also injures the India-rubber, although of course in a less degree. Nitric Acid is used in the treatment of leather cuttings to reduce them to a glutinous mass before being mixed with India-rubber, and is also used in making certain substitutes.

Oil of Vitriol.—See Sulphuric Acid.

Oleic Acid.—An acid found in certain animal and vegetable oils, such as olive oil, sperm oil, etc. It has been used in certain substitutes for hard rubber, like voltit, and by Hunt for recovering waste vulcanized rubber under heat, methylated spirit being added later to precipitate the rubber, which was then washed in weak caustic soda.

Oxalic Acid occurs as transparent, colorless prisms, with a very sour taste, soluble in both cold and hot water. It is produced by either the action of the hydrate of potash, or of nitric acid upon most organic compounds. It is very poisonous. Gutta-percha was cleansed by Lorimer's process by boiling in water mixed with this acid.

Oxalate of Lime.—Quick lime slaked by water in which is oxalic acid is given this name. Used in certain Gutta-percha compounds.

Permanganate of Potash occurs in dark red prisms of a greenish color which, when dissolved in water, gives a purple red. It is a decided oxidizer, and is used as a disinfectant. It is also called chamelon mineral. Used in certain artificial leathers

Peroxide of Hydrogen.—This is a powerful oxidizing agent, largely used as a bleaching agent, and also as an antichlor for use after chlorine bleaching. It comes in the form of a colorless liquid, and has a specific gravity of 1.45. Neither the alkaline nor the acid solutions of this reagent seem to impair vulcanized India-rubber. In certain cases Peroxide of Hydrogen has been used in removing the bloom from rubber, which it does most effectively; besides, it seems to penetrate the surface of the rubber and dissolve the sulphur. It also has a curious effect on colors, brightening some reds wonderfully, dulling others, and rendering whites much whiter. One curious effect that it has upon India-rubber is to bring out any surface imperfections in a marked degree.

PHOSPHATE OF SODA.—A crystaline colorless substance containing 60 per cent. of water, which is given up on heating to 248° F., leaving behind a dry mass. The commercial article frequently contains sulphate of soda as an impurity. The crystals have a specific gravity of 1.5, melt at 95° F., and are readily soluble in water. By long drying at 113° F. the water of crystalization may be entirely driven off. The presence of this material is called for in a certain compound for dental vulcanite, where it is incorporated with rubber, sulphur, and phosphate of lime, the idea being that less sulphur is required than in the ordinary compounds.

PHOSPHORIC ACID.—See Phosphorus.

POTASH.—This substance, a carbonate of potassium, is usually met with commercially in small colorless crystals. It is prepared in a variety of ways and forms, the basis from which is prepared what is called caustic potash. Pearl ash is a crude form of potash mixed with the caustic variety and a sulphuret of potassium. Used in certain proofing compounds where low heat is required for cure. It was used by Charles Hancock, mixed with water in a bath, to improve the quality of Gutta-percha. He found, by boiling the Gutta-percha in such a bath for an hour, that it did not oxidize in the open air as badly. An old-fashioned process for treating unvulcanized thread was to steep it in a hot solution of carbonate of potash, which greatly increased its strength. (See Caustic Potash.)

QUICK LIME is the impure oxide of calcium obtained by heating or burning chalk, marble, or limestone, or any carbonate of calcium. Its well-known attraction for water renders it unstable but also valuable where dying qualities are desired. Blizzard claimed to be able to make a perfectly transparent rubber by treating it with soda and water, in which was a little Quick Lime.

RENNET is made from the inner lining of the true stomach of the sucking calf and gets its value from the gastric juice contained thereni. The membrane, after treatment, is salted and stretched out to dry. It is advised in the Vaughn process for coagulating Balata.

SALICYLIC ACID is obtained from the creeping plant known as wintergreen. It is prepared from the oil of wintergreen (oil of

Gaultheria), which is distilled in large quantities in Luzerne county, Pa. It is soluble in the following proportions: 1 part of the acid dissolves in 450 of water, or 2.4 of alcohol. It melts at 312° to 314° F. Salicylic Acid was used in an artificial leather compound for reducing leather dust to a paste, after which it was mixed with glue under heat, and treated to an alkaline solution.

Sal Ammoniac.—See Chloride of Ammonium.

Salt.—See Chloride of Sodium.

Saltpeter is niter or potassium nitrate. It is a crystaline substance, white, and having a saline taste, and is a very strong oxidizer. It is used in the manufacture of artificial elaterite. In Gridley's process for recovering rubber, by exposing it to flame, saltpeter was added to remove the smell.

Saleratus.—See Carbonate of Soda.

Sal Soda.—See Carbonate of Soda.

Soda.—See Carbonate of Soda.

Sodium Hyposulphite.—A 1 per cent. solution is used for removing traces of chlorine where its presence is suspected in rubber.

Soluble Glass (known also as waterglass) is a silicate of soda or potash. It is usually sold in solutions of varying density, the commonest being 33° and 66°, by which is meant that the solution contains either one third or two thirds solid waterglass. Acids readily precipitate the silica from these solutions as a gelatinous mass. It is used in certain shower-proof compounds and in compounds of the Algin (which see) type.

Stearic Acid.—See Stearine.

Sulphate of Alumina.—The active principle of alum. Often sold as concentrated alum. Occurs commercially as white square cakes, somewhat transparent, and capable of being cut with a knife. Readily soluble in water, and contains a small quantity of free sulphuric acid, potassa, and soda alum. Its specific gravity is about 4; water of crystalization 48 per cent. Its composition indicates a usefulness in compounding sponge rubbers. Used in linseed oil compounds, for wagon covers. (See Alum.)

Sugar of Lead.—This is used in certain rainproof compounds, one of which is 16 parts of compounded rubber, 128 parts of paraffine wax, 1 part of Sugar of Lead, 1 part of alum in powder. India-rubber compound contains no sulphur. Used also in artificial rubber and artificial ivory. (See Acetate of Lead.)

SULPHATE OF COPPER.—Sometimes called blue or Cyprus vitriol. Occurs in commerce in masses of large blue crystals having a specific gravity of 2.28, and containing 36 per cent. of water of crystalization, and a varying additional percentage of entangled moisture. Heated for some time at 212° F. all the entangled water may be driven off, together with four-fifths of the water of crystalization, the residue being a bluish white powder. Sulphate of Copper is used in attaching rubber to iron during vulcanization.

SULPHATE OF SODA occurs commercially in colorless crystals which deteriorate in contact with the air, and hence should be kept in well closed vessels. It contains a very large amount—nearly 60 per cent.—of water of crystalization, which is yielded on heating to 302° F. Its reaction is alkaline. Sulphate of Soda was used by Hancock in vulcanizing Gutta-percha.

SILICATE OF SODA.—See Soluble Glass.

SOAPS.—Various kinds of soaps are used in rubber manufacture. Pure Castile soap, for instance, is dissolved in rain water and made into a soft soap that is used to "slick" molds that the rubber, during vulcanization, may not adhere to them. Some manufacturers use by preference white soda soap made from caustic soda and olive oil. Resin soaps are also used in certain shower-proof compounds. A further use for soap is in the manufacture of water varnishes for luster coats and blankets. A soap compound for wagon covers is made of 50 pounds of soap dissolved in 15 gallons of water, heated to 250° F., to which is added 25 pounds of sulphide of zinc. A half pint of rubber dissolved in olive oil by heat is added to each gallon of the above mixture. Whiting, lampblack, or coloring matters may be added. Vulcanized rubber, beeswax, resin oil, argillaceous earth, and alkaline soap form the basis of Sorel's substitute for rubber.

SULPHURIC ACID (called also Oil of Vitriol), when pure, is a colorless oily looking heavy liquid of a sharp, sour taste. It is very corrosive, and has a great attraction for water; hence wood and other organic bodies are charred by its depriving them of their water. The specific gravity of the commercial acid is usually about 1.83, or 66° Beaume, containing 94 per cent. of acid. Sulphuric Acid is used in the coagulation of Madagascar rubber. The Orinoco Co. are also said to coagulate India-rubber by mixing the milk of the *Hevea* with sulphuric and carbolic acid. Commercial Sulphuric Acid is said to coagulate 55 times its volume of gum, while the carbolic acid acts as an antiseptic in the juice, improving its keeping qualities. It is a question whether rubber treated this way is as good as that obtained by the smoking process. Rubber immersed in Sulphuric Acid at 140° F. remained a month and came out stronger, apparently, than when it went in. Sulphuric Acid is used in paste blacking, mixed with boneblack, vinegar, molasses, and caoutchouc oil. Concentrated Sulphuric Acid colors Gutta-percha brown, throwing off at the same time Sulphuric Acid fumes. Nevertheless, a paste of this acid and charcoal was added by Hancock to Gutta-percha to make it pliable. Sulphuric Acid may be expected to attack vulcanized rubber compounds in which there are large proportions of chalk, lead oxides, or barytes. Sulphuric Acid is very largely used in destroying the fiber found in ground waste rubber; indeed it is the basis of what is known as the acid reclaiming process. When thus used the acid turns whiting into sulphate of lime.

TANNIC ACID.—See Tannin.

TANNIN includes a number of substances, some of which are crystaline and others amorphous, with a marked astringent taste, and no smell. The solutions are acid, soluble in water and alcohol, and yields precipitates with most metallic oxides. It is the active principle of oak bark, hemlock bark, catechu, and many other materials usually used for tanning hides. Pure Tannin is a light powder of a yellow greenish hue, soluble in water, alcohol, and ether. Its solution precipitates glue. It is used with sulphate of alumina, waterglass, and glue in shower-proofing. Tannin has been claimed to be injurious to rubber, the reason being that rubber thread used in gorings is often destroyed at points close to its junction with the leather. It is more likely, however, that it is the oil or oleic acid that effects the destruction. Tannin was largely employed by Austin G. Day in many of his "kerite" compounds with excellent effect. It is also used in the manufacture of certain puncture fluids, together with glue and glycerine.

TARTARIC ACID is found usually in the form of transparent colorless prisms, which have an agreeable acid taste, are not affected by the action of the atmosphere, and are soluble in either alcohol or water. Nitric acid or peroxide of lead act upon Tartaric Acid, turning it into formic and carbonic acid. This acid is very abundant in the vegetable kingdom, being found in many fruits. Used under Vaughn's patent in coagulating Balata. Vulcanized rubber immersed in Tartaric Acid at 140° F. remained a month, and was apparently unharmed.

TUNGSTATE OF AMMONIA.—A crystaline body which is very soluble in water and becomes covered with a white bloom on exposure to the air. Used with boracic acid, kauri, borax, and rubber in the production of the woodite fireproof compositions.

TUNGSTATE OF SODA.—Prepared commercially from wolfram and soda ash; usually contains about 14 per cent. water of crystalization; and is in the form of colorless crystals. Mixed with a solvent such as methylated ether, it is added to soluble gun cotton, castor oil, and gum copal, forming a substitute for India-rubber.

TUNGSTIC ACID is derived chiefly from wolfram, which is a tungstate of iron and manganese. Tungstic Acid is analogous to sulphuric and chromic acid. It has been used in connection with paraffine, gelatine, and metallic oxides in proofing compounds.

CHAPTER X.

VEGETABLE, MINERAL, AND ANIMAL OILS USED IN RUBBER COMPOUNDS AND SOLUTIONS.

THE use of oils in the rubber manufacture has kept pace fully with the use of gums, substitutes, and reclaimed rubber. The addition of earthy or metallic or vegetable ingredients in dry mixing has rendered many a good rubber somewhat intractable—a fault which the right oil has often rectified. As a rule, vegetable oils are chosen, as they are rarely harmful to the gum. Many mineral oils are also freely incorporated in certain compounds. Animal oils have always been viewed with more or less suspicion, however, and with good reason, for manufacturers have constantly before them rubber goods that have lost their life and elasticity through contact with lubricants made of such oils and fats. Nevertheless certain of them may be and are used. The essential or volatile oils are used to a certain extent in rubber manufacture. These oils, as a rule, are liquids which give the peculiar odors of plants from which they are derived. Their use in rubber is to impart to it a pleasing odor.

ALUMINUM LANOLATE.—This is a product of French wool grease (which see), made by adding a solution of alum. After the addition of the alum, it falls in a brown precipitate. It is then dissolved in mineral oil, forming a jelly-like mass which is said to compound readily with either India-rubber or Gutta-percha, and is soluble in any of their solvents. It is possible that this may have some both softening and preservative influences on India-rubber, as is claimed, but it should be used with considerable caution.

ANHYDROUS PARAFFINE OIL.—Water-free paraffine oil (which see.)

BIRCH OIL.—The fine white bark of the birch tree yields a red oil, nearly one-fourth of which consists of the sap *phenol,* which gives the well-known odor to Russia leather. The residue, or green part of the birch, yields neither acid nor alkaloid, and forms with alcohol a fluid solution which, when once dried, is unacted on by alcohol. It is chiefly obtained from northern

Europe and Siberia, and has recently been made also in Germany and Austria, where it is known as Jackten oil. This substance will unite with the most brilliant colors, and has been used in France for waterproofing textile fabrics. In connection with shellac, resin, and aniline, it is used in the form of a substitute for Gutta-percha in insulation.

Blown Oils.—These are prepared by heating fixed oils in a jacketed kettle and blowing a current of air through the fluid. Under this treatment, oils become much more dense and also viscous; indeed, in many physical aspects, they resemble castor oil, but differ in that they can be mixed with mineral oils and as a rule are not easily soluble in alcohol. Blown oils made from linseed oil, rape oil, poppyseed oil, and cottonseed oil are sometimes used in the manufacture of rubber substitutes instead of the raw oils. Known also as Thickened Oils, Base Oils, Soluble Castor Oil, etc.

Bone Oil is obtained by the distillation of animal gelatinous substances, principally in the calcining of bones for the preparation of boneblack. Its specific gravity is 0.97. It is sometimes called Dippel's Oil (which see.)

Camphor Oil.—A liquid of a light reddish brown with a yellowish tint, a strong odor like camphor, and a bitter camphor-like taste. Its specific gravity is 0.94. Japanese oil varies in color from colorless through pale straw, yellow, to black, and has a specific gravity of 0.898 for the colorless to 0.99 for the very dark. This oil is used in the manufacture of celluloid varnishes, paints, lampblacks, etc. It is used also as an adulterant for such oils as sassafras oil. It is one of the best solvents for resins, and dissolves 46 per cent. of rosin, 9 per cent. copal, and 35 per cent. of mastic.

Caoutchouc Oil.—Made by digesting 55 parts of India-rubber in 450 parts of linseed oil. The only large use for this oil is in Germany, particularly in the army, where it was used for coating various articles to prevent their rusting. The following substances are found in Oil of Caoutchouc: Eupoine, butylene, caoutchoucine, isoprene, caoutchine, and heveene.

Castor Oil.—A colorless or pale greenish transparent oil, very viscous and thickening on exposure to the air. It has the highest specific gravity of any known natural fatty oil—0.958. It is adulterated frequently with resin oil and rape, linseed, and cottonseed oils, especially the "blown" variety. Used in cheap proofings without rubber with Kauri gum; also in collodion and rubber proofing. It is used in the production of substitutes like gum fibrine, and also with chloride of sulphur in producing amber colored substitute.

CHOLESTERIN.—See Lanichol.

COD OIL or COD-LIVER OIL is obtained from the livers of codfish. Newfoundland and Norway are the principal manufacturing points. The finest is a very pale, clear, golden yellow, the color deepening to a brown in the second and third grades. Its specific gravity is 0.923 to 0.929. One part of oil is soluble in from 40 to 20 parts cold alcohol, or 30 to 17 parts hot alcohol. The lower grades are the more soluble. It is much adulterated. Is compounded with India-rubber, beeswax, linseed oil, litharge, and asphalt as a waterproofing for leather and with India-rubber, beeswax, and turpentine as a dressing for hides.

COLZA OIL.—See Rape Oil.

CORN OIL (also known as Maize Oil).—Made from the seed of Indian corn, the plant being known botanically as *Zea mays.* There are two processes of manufacture: (1) in which the seed is pressed before it is used for the manufacture of starch, which produces oil of a golden yellow color, and (2) where it is recovered from the residue of the fermentation vats where the corn has been used in the production of alcohol. This oil is dissolved sparingly in alcohol, but very readily in acetone. The oil is almost without drying powers. Neither boiling nor the addition of lead when boiling gives it definite drying properties. If it is heated, however, and a current of air passed through it, and manganese borate mingled with it, it dries after a fashion. It is largely used at present in the manufacture of what are known as Corn-oil substitutes.

CONSOLIDATED OIL.—See Stearine.

COTTONSEED OIL is made from the seeds of the cotton plant, usually the *Gossypium herbaceum.* The crude is of a ruby red almost black color. The refined is pale yellow and possesses a pleasant nutty taste. It is a semi-drying oil, and is rarely adulterated except when linseed oil is very cheap. On standing it deposits stearine in waxy flakes. Much used in making substitutes for rubber. It is also used in the production of artificial elaterite, and with paraffine oil for canvas proofing. For Cottonseed blown oils see Blown Oils.

CREOSOTE OIL is a distillate from wood tar. It is an oily liquid with a smoky taste, and is antiseptic. It should be colorless but is usually yellow or brown, due to impurities or to exposure. The best is made from the beech. A similar oil is distilled from coal tar. Mixed with red oxide of mercury it has been used to coat the fabric of which cotton hose is made as a preservative; with India-rubber and sulphur it has also formed an insulating compound for telegraph wires. It is used in some rubber works where it is arranged that the fumes of the naphtha are carried off into it, which it rapidly absorbs, to be later recovered by distillation.

EUCALIPTIA.—A fragrant, refreshing volatile oil, twenty to forty times as strong a disinfectant as fluid carbolic acid. It is prepared from eucalyptus oil.

EUCALYPTUS OIL.—An aromatic oil found in the leaves of the *Eucalyptus globulus*, in Australia. The odor of the oil is extremely pleasant, smelling not unlike oil of verbena. This oil is said to be most advantageous, used in small quantities in connection with solvents for India-rubber, as it tends greatly to accelerate complete solution. It also breaks down refractory samples of the gum and renders all of the compound homogeneous. It is said that one-third of the time may be saved if from 4 to 6 per cent. of this oil is used in the solvent. It is especially good for low-grade gums. It has also great solvent power on all resins and gums, including India-rubber and Gutta-percha. With the addition of a little methylated spirit it will dissolve even Kauri gum, cold. It is also used in dissolving asphalt for photograph varnish.

ESSENCE OF PETROLEUM.—Obtained during the refining of Petroleum, and known also as petrolatum, vaseline, petroleum jelly, etc. (See Vaseline.)

FISH OIL.—Obtained from all parts of the bodies of common fish by boiling. Fish whose livers yield oil commercially do not give fish oil, and those bodies that yield oil, do not give liver oils. Principally prepared from Menhaden. Its specific gravity varies betwen .915 and .930. Fish Oil is used in the manufacture of the substitute known as volenite. It is used, however, only as a vehicle for carrying resin into the fiber, being afterwards wholly removed.

FRENCH WOOL GREASE.—See Lanoline.

GLYCERINE.—A clear liquid of oily consistency and sweet taste, without odor. When pure it has a specific gravity of 1.26. The Glycerine of commerce is a by-product of the soap manufacture, chemical reaction occurring when the fat is treated with a caustic alkali, giving rise to a compound of a fatty acid and alkali to form a soap, while the Glycerine is at the same time liberated and goes into solution. Glycerine is not acted upon by oxygen, and therefore more closely resembles mineral oils, such as are used in rubber mixing, than it does the drying oils that go to make up substitutes. It has absolutely no solvent action on rubber.

A recent German patent calls for the addition of Glycerine because of its oil resisting qualities. In the compound used are 6 pounds of rubber, and 1 pound of Glycerine, together with whiting, litharge, and sulphur. A soap made of Glycerine and an alkaline fluid is also used as a cleansing and polishing medium in the last stages of the manufacture of certain cut sheet goods. Glycerine combined with gelatine and borax has been used as a wash for both black and red rubber surfaces.

Glycerine was the basis of a well-known deodorizing composition for India-rubber, the other ingredients being of an alkaline nature. A bath of Glycerine has also been used for experimental work in vulcanizing India-rubber, and also for rubber stamp making. In this kind of work, the mold and its contents are immersed in the Glycerine so that the liquid just covers the top of the mold; heat is then applied to the Glycerine, and the mold in turn becomes hot and the rubber vulcanizes. It is also used to a certain extent in good grades of white rubber, as it gives a softened effect to the compound. Glycerine, in connection with glue, gelatine, molasses, and tannin, is used in the manufacture of puncture fluids for tires. It is also used in clothing compounds, and in cellulose products like pegamoid. Used in rubber, a little of it increases the resiliency of the product. Another use for Glycerine is to prevent fabrics from mildewing. The fabric is coated with it before being frictioned.

JAPAN WAX.—A white or pale yellow vegetable fat, with a specific gravity of 0.97 to 0.98. It is used in wax matches, candles, and for adulterating beeswax. A special use for it, that has arisen within the last few years, is in the manufacture of cravenette cloths.

LALLEMANTIA OIL is obtained from the seeds of the *Lallemantia iberica,* a plant cultivated in Russia. This is one of the best drying oils, being said to surpass even linseed oil, but its chief use is for illuminating purposes. In Europe it is said to have been used instead of linseed oil in rubber substitutes.

LANICHOL.—A product of lanoline (which see), made from the oil of sheep's wool. It combines with Gutta-percha and India-rubber in any proportion to a perfectly homogeneous mass. This grease does not oxidize and is wholly antiseptic. It has no smell, and is impervious to the action of alkalies or to dilute sulphuric acid. It is said that, used in connection with Gutta-percha, the melting point is considerably raised, while it does not diminish the insulating property. An insulating compound given is 50 parts by weight of Gutta-percha, 30 parts of India-rubber, 20 parts Lanichol. The inventor claims that it renders Gutta-percha less liable to oxidation, improves its elasticity and tenacity, and diminishes its liability to become sticky. Patented in the United States and Great Britain by Robert Hutchinson.

LANOLINE is also known as wool grease, recovered grease, and brown grease. It is the natural grease found in sheep's wool and recovered from it while the raw wool is being prepared for spinning. A similar grease, made from scoured woven goods, is known as Yorkshire grease. It is a thick yellow or brown offensive smelling greasy paste. Commercial Lanoline is lighter colored and consists of about 80 per cent. of pure wool fat and 20 per cent. of water. It possesses in a remarkable degree the property of taking up water without losing its vaseline-like consistency. Is largely used in ointments.

Lanoline, mixed with India-rubber, works up into an exceedingly sticky mass, and is used as a medicinal plaster. It is said that, while it possesses the adhesive properties of the regular plaster, Lanoline takes up the medicament, and while very sticky can be readily removed from the skin. It is used for the purpose of softening India-rubber, and was advised for use in tires, as it was said to soften the compound, and to keep the tire from decay, and from consequent surface cracking. It was also said to be used in boot and shoe work.

LARD OIL is prepared by the cold pressing of lard, which, of course, is the fat of the hog. It is a colorless, limpid liquid, although poorer grades are brown. Its specific gravity is 0.915. It is frequently adulterated with rape oil and cottonseed oil. Lard Oil, mixed with powdered pumice stone into a thick paste, is used for polishing hard rubber.

LINSEED OIL is pressed from the seeds of the flax plant *(Linum usitatissimum)*, grown chiefly in India and Russia. The trade recognizes two qualities of Russian seed—yielding the Black sea Linseed Oil, and the Baltic Linseed Oil—while that coming from India is known as East India oil. Of these, the Baltic is the best, and the East Indian the poorest in quality. The two lower grades are not up in quality for the reason that the Black sea seed contains a certain amount of hemp-seed, while that from India is usually mixed with rape, cameline, and mustard seeds. The oil which is expressed from these seeds is of a golden yellow color, with a peculiar taste and odor. Linseed Oil becomes easily rancid in the open air, but when spread in thin films dries into an insoluble substance which has been called linoxyn. Linseed Oil is adulterated sometimes by fish or mineral oils, and by resin oils. Old tanked Linseed Oil is used in the preparation of what is known as boiled oil; that is, it is heated in a high temperature that it may more rapidly dry when used in varnish. This drying process is hastened by the addition of manganese dioxide, litharge, etc. Boiled Linseed Oil is much darker than raw oil, having a brown red shade. It is also much more viscous and has a higher specific gravity. Boiled oil is adulterated in the same manner as is raw Linseed Oil, the adulterants being resin oils, resin, and mineral oils.

In rubber compounding Linseed Oil is very often used. A

very simple formula for waterproofing canvas is India-rubber, litharge, sulphur, and Linseed Oil. It is also used in rubber varnishes, to a certain extent in molded goods, and quite largely in hard rubber compounding. It is used in the manufacture of rubber substitutes, and is well known as it is the basis of a great many of the vulcanized oil substitutes. Linseed Oil that is intended for mixing in linoleum is exposed to the air until it is thoroughly oxygenated. In this state it is insoluble in alcohol, chloroform, ether, and ordinary solvents.

Lithographic Varnish.—This is obtained by boiling linseed oil at a temperature higher than that at which boiled oil is prepared, nor are dryers added during the boiling. It is a perpectly clear, transparent substance, the best quality being nearly as light as raw linseed oil. There are two ordinary grades of Lithographic Varnish. One is known as "burnt oil," which is obtained by bringing raw linseed oil up to its flash point, and allowing it to burn until the required thickness is reached, it being constantly stirred meanwhile. "Oxygenated oil" is a linseed oil varnish made by treating the oil with oxygen in jacketed kettles, heated by steam. The product is as light colored as raw linseed oil, but heavier. It is also more readily soluble in alcohol, and has marked drying powers.

Mirbane Oil.—See Nitrobenzene.

Manganated Linseed Oil is used in certain rubber compounds where more of a drying effect is needed than is found in the raw linseed oil. It is linseed oil that has been boiled with peroxide of manganese to increase its drying qualities. (See Boiled Oil.)

Mustard Oil.—Black Mustard Oil is obtained from the seeds of the *Sinapsis nigra.* It possesses a mild taste, is of a brownish yellow color, and in its chemical composition closely resembles rapeseed oil. It is a by product and is largely used in soap making. White Mustard Oil is made from the seeds of the *Sinapis alba.* It is of a yellow color, and is almost identical with black Mustard Oil. Both of these oils have been used in the manufacture of rubber substitutes.

Neatsfoot Oil.—A pale, yellow, colorless oil, obtained from the feet of oxen by boiling in water. It has a smooth pleasant taste. On standing it deposits stearine. It is largely adulterated with cheaper animal or vegetable and even mineral oils. Neats-foot Oil, mixed with Gutta-percha, tallow, sweet oil, and oil of thyme, is used as a rust preventative. It is used in connection with beeswax, India-rubber, and Burgundy pitch in a composition for dressing leathers or hides.

NITROBENZENE (also called "oil of mirbane" and "imitation oil of bitter almonds") is a yellow aromatic liquid produced by the action of nitric acid on benzene. It is used in perfumery and turned out in great quantities during the manufacture of anilines. It is used also in certain insulating compounds in connection with asbestos, powdered glass, vulcanized rubber, castor oil, resin oil, and celluloid in solution.

OIL OF LAVENDER has no perfume when new, but develops it on being exposed to the air. It is distilled from the flowers of the *Lavandula vera,* and is used sometimes to deodorize rubber goods.

OIL OF LEMON is obtained from fresh lemon peel. A very volatile yellow or colorless oil; specific gravity 0.858; soluble in bisulphide of carbon, and absolute alcohol; often adulterated with fixed oils and alcohol; dissolves sulphur, phosphorus, resin, and fats; used to deodorize certain proofing compounds, cologne sometimes taking its place.

OIL OF ORRIS, or ORRIS OIL, is found commercially and is prepared from the root. It is lighter than water, and of the consistency of butter. Melts at 100° F., and is miscible with alcohol. Its odor is like that of violets. Is used in rubber as a deodorizer.

OIL OF PEPPERMINT.—A greenish yellow colorless oil, becoming reddish with age; of a strong and aromatic odor; and warm, camphor like, very pungent taste; specific gravity from 0.902 to 0.920; used in fine goods for its odor.

OIL OF ROSEMARY.—An essential oil of the specific gravity 0.896. Colorless and having the odor of rosemary. Used with India-rubber, paraffine, and spermaceti in waterproofing compounds, and, where rubber is present, to neutralize its odor.

OIL OF TAR.—An oil distilled from tar. It is a mixture of several lighter oils, and is made up of liquid hydrocarbons which hold in solution small quantities of anthracine, naphthaline, and paraffine. It is sometimes used for mixing with lubricating oils, and for coating bags that are to hold alkaline earths, the interior of the bag being washed with chloride of lime. The Earl of Dundonald recommended Oil of Tar as a coating for rubber, claiming that it had a preservative effect. It is also used in compounds for surface clothing.

Oil of Thyme (also called Origanum Oil) is extracted from the flowers and leaves of the *Thymus vulgaris.* It is yellowish red in color; its specific gravity is 0.92; and it has a pungent taste; it is used to disguise the odor of ale cements.

Oil of Wormwood.—A pungent essential oil distilled from the *Artemisia absinthium;* employed at an early day to deodorize spirits of turpentine when used in rubber.

Oleargum.—A black viscid liquid of an oily nature used as a dull finish wash for rubber boots. Its composition is a trade secret.

Oleum Succini.—The same as Oil of Amber (which see); used in the manufacture of soap substitutes.

Olive Oil is expressed from the fruit of the olive tree, principally in the countries of Europe bordering on the Mediterranean. Its specific gravity is 0.916. It is adulterated frequently with cottonseed oil. Olive Oil is used in taking impressions from type-faces in the matrix in which rubber type is cured. Mayall suggested the mixing of Olive Oil with clay until it formed a soft putty, and then incorporating it with the India-rubber, the proportion being ¼ pound of oil to 30 pounds of gum. The use of the oil enabled the goods to be more largely adulterated; he also used Olive Oil in connection with devulcanized rubber, not as a solvent, but because he claimed that it combined with the gum and improved its quality. Olive Oil is also used in hard rubber compounding. Rubber is sometimes heated up in Olive Oil mixed with zinc, soap, and borax for a proofing solution. It is also used in the manufacture of pegamoid.

Palm Oil is obtained from the fruit of various species of palm, principally from the west coast of Africa, and is known in commerce under as many names as there are ports of shipment. It is expressed in a very rough fashion by the natives, who stir the palm kernels in holes in the ground until fermentation sets in and the oil rises to the surface. They also sometimes press the oil from the fresh fruits. The harder grades of Palm Oil are yielded by the former procees, the latter giving the finer oils. Palm Oil varies in consistency. Its specific gravity is 0.945; its color yellow to reddish; its odor that of violets. It yields a soap readily with alkalies and dissolves in ether and in alcohol of 0.848 specific gravity. Palm Oil is very rarely adulterated, unless it is done by the native gatherers, who sometimes add sand as a make-weight. Commercially, where sand and water together exceed 2 per cent., an allowance is claimed from the seller.

White Palm Oil is that which has been bleached by heated chemicals or exposure to the air. "Lagos oil" has about the same consistency as butter, while "Congo oil" is as thick as tallow. Palm Oil is used largely in the manufacture of mechanical and dry-heat goods, chiefly to enable dry ingredients to mix more easily with India-rubber. It has also been used in the recovery of waste rubber by the mixing of the finely ground rubber with it and exposing the mass to a heat of 572° F. Palm Oil residuum is used in connection with resin oil as an insulator. Palm Oil is also used in the production of artificial elaterite.

PARAFFINE OIL is a petroleum product; it is also prepared from coal tar and wood tar. It is a waxy substance of a white color, much resembling spermaceti. It is used chiefly as a lubricant, and is not acted upon by most of the chemical reagents. Paraffine Oil mixed with cottonseed oil is used in certain canvas proofings.

PETROLEUM OIL (also known as Rock Oil) is a dark, ill smelling liquid, obtained from wells sunk in oil-bearing sands. Some Russian oils, however, are colorless. White Rangoon oil contains so much paraffine as to have the consistency of butter. The specific gravity of American petroleum varies from 0.8 to 0.85 or 0.9.

PETROLEUM PARAFFINE.—See Vaseline.

PETROLEUM JELLY.—See Vaseline.

PETROLATUM.—See Vaseline.

POPPYSEED OIL is obtained by pressing the seeds of the common poppy *(Papaver somniferum)*. Commercially there are two grades: (1) white Poppyseed Oil and (2) red Poppyseed Oil.

This oil has a pleasant taste and no odor; it is rarely adulterated with other oils, although occasionally sesame oil is found in it; it is an excellent drying oil, and its lower grades are used in the manufacture of soaps; its use in the rubber industry is chiefly in the manufacture of substitutes.

RAPESEED OIL (also know as Colza Oil) is a pale yellow in color, with an unpleasant harsh taste. Its specific gravity is about 0.916. It is largely adulterated with both vegetable, mineral, and fish oils. It is obtained from the seeds of the *Brassica campestris*, and of several varieties of this genus which are cultivated. American oils from all of these are termed colza oil, or rape oil indiscriminately. In Europe, however, rape is one kind of oil and colza is another. There is also what is called the summer oil and the winter oil, a distinction which is of no interest to rubber manufacturers. Rape oil is hardly a semi-drying oil, nor is it yet a non-drying oil, but about half way between the two. It is used in the manufacture of certain rubber substitutes. Mixed with India-rubber it has been used as a somewhat costly mixture for lubricating machinery.

ROSIN OIL.—Made by subjecting resin to destructive distillation. The resultant oil is heavier than mineral oils, and its chemical composition is quite involved. It is largely made up, however, of hydrocarbons, with a certain amount of resin acids. Used in making a waterproof solution, by the addition of Japan wax and gum thus, in the manufacture of a solution for treating hides and leather. Used also in compounds for calking ships in which India-rubber has a part, and is an important ingredient in the manufacture of guttaline.

RUSSIAN MINERAL OIL.—Petroleum from the Baku oil wells in Russia.

SHALE OIL.—Chiefly produced in Scotland from a dark, coal-like looking material called shale. It is similar in nearly all respects to petroleum oil. Used with asphaltum in certain insulating compounds.

SLUDGE.—The brown or black residue obtained in the refining of petroleum after all the lighter oils have been distilled off. Known also as Petroleum Residuum. (See Sludge-oil Resin.)

STEARINE.—An important ingredient in animal and vegetable fats. It is quite solid, and increases the hardness, and raises the melting point of fat. Commercially, Stearine is also known as stearic acid. It is an important element in the manufacture of cravenettes, where it is used with ozocerite, beeswax, paraffine, and Japan wax.

TALLOW.—Beef tallow, when fresh, is almost white, free from disagreeable odor, and almost tasteless. On the other hand, foreign tallow runs from white to yellow and is often quite rancid. Tallow is often adulterated with resin oil, cocoanut oil, cottonseed oil, and paraffine wax. It is used in non-drying cements in connection with slaked lime and India-rubber. In connection with India-rubber it is also used in the production of what was known as Derry's waterproof harness oil, which was made of India-rubber, Tallow, seal oil, and ivory black. An etching varnish is made of Gutta-percha, turpentine, beeswax, and Tallow. A small amount of this was used by Hancock in compounding for softening Gutta-percha. It is used with Gutta-percha in shoemakers' wax, and also in certain proofing compounds with India-rubber, pitch, and linseed oil. Mixed with India-rubber, beeswax, and linseed oil, Tallow makes an excellent dressing for leather.

TURPENTINE was used in one of the earliest formulas in the manufacture of devulcanized rubber. (See Spirits of Turpentine.)

VASELINE is the purified residue from the distillation of petroleum. Its specific gravity is .875 to .945. It is insoluble in water, barely soluble in cold, but soluble in boiling absolute alcohol, and in ether, bisulphide of carbon, oil of turpentine, benzine, and benzol. It is the basis of a cheap waterproofing process, the other ingredients being silicate of soda, alum, and hot water. Vaseline is used quite often in general compounding for its softening effects. It is also combined with menthol and gum alibanum in the manufacture of porous plasters. Vaseline has been used in the manufacture of substitutes similar to ruberite.

VULCANIZED OIL.—See Rubber Substitutes.

WALNUT OIL.—Cold drawn oil is very fluid, almost colorless, and of an agreeable nutty flavor. Hot pressed oil has a greenish tint and an acrid taste and smell. Is used in rubber substitutes, particularly in those in which peroxide of lead appears as a dryer.

WHITE DRYING OIL.—Bleached linseed oil.

CHAPTER XI.

SOLVENTS USED IN INDIA-RUBBER PROOFING AND CEMENTING AND IN COMMERCIAL CEMENTS.

THE beginnings of the manufacture of India-rubber consisted in putting the gum in solution; and it was a considerable time before the discovery of the present processes of dry mixing, which are employed in the production of the greater part of the rubber goods now made. There are certain lines, however, where the use of solvents is still both necessary and economical. In the mackintosh manufacture, for instance, the rubber is in almost every instance spread in the form of solution, as a thinner coat can be spread in this way, offsetting the cost of the solvent. Many sheetings in various colors that, only a few years ago, were calendered, are now coated by the means of solution. In the making up of almost all lines of rubber goods, certain cements are necessary, and these are ordinarily made in the factory that produces the goods. The cements that are sold in bulk, such as channeling cements, for leather shoe manufacturing, as well as cements that are sold in smaller packages to repair men in the cycle industry, all consist of rubber and analogous gums treated with some suitable solvent. Before discussing the ordinary and the extraordinary solvents that interest the rubber manufacturer, it may be well to consider what the various solvents can do.

The following tables showing the solubility of India-rubber are of exceeding interest, therefore. The first, which is taken from the *Journal of the Society of Chemical Industry*, is a table of the solubility of masticated caoutchouc in solvents:

100 parts of:	Ceara Rubber.	Para Negroheads.	Sierra Leone Rubber.
Ethyl ether	2.6	3.6	4.6
Turpentine	4.5	5.0	4.6
Chloroform	3.0	3.7	3.0
Petroleum benzene	4.4	5.0	{4.0 / 4.7
Carbon bisulphide	0.4	None.	None.

Hoffer gives, as a result of his individual experiments, the following table of solutions, the samples in each case being 100 parts of well-dried India-rubber:

In bisulphide of carbon	65 to 70
In benzol	48 to 52
In oil of turpentine	50 to 52
In caoutchine	53 to 55
In ether	60 to 68
In camphene	53 to 58

The great differences between various grades of rubber have been found to be due, as much as anything, to the amounts of resins that are to be found in them. As these resins are soluble, and in some cases can be removed, it is important that rubber manufacturers not only appreciate their presence, but, where it is practicable, dissolve them out. These resins, according to Lasselles-Scott, who furnishes the following valuable table, consist of abietic acid or some other similar body:

Description of Rubber.	Normal Resin (soluble in 85 p. c. Alcohol).	Description of Rubber.	Normal Resin (soluble in 85 p. c. Alcohol)
Para	.91	Ceara	1.16
Para	.60	Assam	6.45
Para	1.62	Assam	4.88
Para	1.14	Burma	5.20
Para	.85	Rio	3.37
Madagascar	4.06	Africa (various)	8.23
Madagascar	5.22	Africa (various)	10.60
Madagascar	2.84	Africa (various)	6.71
Colombia	3.40	Mangabeira	8.43
Colombia	2.11	Origin unknown	11.14
Ceara	2.33	Origin unknown	7.27
Ceara	1.80	Origin unknown	16.56

In some of them oxygen is a component part, and they are all soluble in alcohol of 85 per cent. strength and upwards. It will be noticed from this table that Para rubber has the least percentage of resin, and, of course, is the most valuable. The samples containing the largest proportions of resin were unmistakably adulterated with other gums during collection.

C. O. Weber gives the percentages of resin in a number of samples of rubber as follows:

Grade of Rubber.	Per Cent. Resin.	Grade of Rubber.	Per Cent. Resin.
Para (fine)	1.3	Sierra Leone	9.7
Ceara	2.1	Assam	11.3
Colombian	3.8	Mangabeira	13.1
Mozambique	3.2	African ball No. 1	22.8
Rio Janeiro	5.2	African ball No. 2	26.1
Madagascar	8.2	African flake	63.9

ACETONE is a colorless mobile liquid, with a very unpleasant taste and peculiar odor, and outwardly resembling alcohol. It is a good solvent for organic substances, and for many gums and resins. When recovered from wood spirit, it is distilled from the calcium chloride compound, generally with methyl alcohol. It has a specific gravity of 0.802. Acetone is the solvent used in the preparation of linoxin.

ALCOHOL, when pure, is a colorless, thin, mobile liquid, of a somewhat disagreeable smell, burning taste, and specific gravity 0.792. What is known as absolute alcohol is that which has been deprived of all water. Its specific gravity is 0.795. It eagerly absorbs water, and, as it becomes more dilute, its specific gravity rises; alcohol of 60 per cent. has a specific gravity of .883. There are a number of forms of alcohol used in the arts. Methylated spirit is a form having the lowest boiling point of the group of alcohols; rectified spirit is a term for alcohol of 95 per cent. and specific gravity .806; fusel oil is a complex mixture of alcohol and various ethers, being a colorless liquid of burning acrid taste and odor very irritating to the lungs, with a specific gravity of 0.818. The last is made usually from potatoes. None of these really are solvents of rubber, but are frequently and largely used in varnishes. India-rubber, when treated with large quantities of alcohol, is deposited in a spongy form, the foreign ingredients in the gum going into solution. Treated in this way it can be made an exceedingly white mass. It is also used in treating many of the pseudo guttas to dissolve out the brittle resinous matters. It has also been claimed that the washing of raw rubber with alcohol dissolves resinous ingredients which are better absent, and that the rubber as a result lasts longer. Rectified spirit is what is generally known, or rather, used, in connection with India-rubber. It is used by the gatherers to coagulate the sap of the Balata, and is used also in the production of resinolines (which see). One of the early uses was to mix with it various solvents—for instance, with spirits of turpentine, coal oil, bisulphide of carbon, ether, chloroform, etc. When ill-smelling solvents were used, it was also often incorporated to neutralize the odor. In the Azo process for reclaiming rubber, 20 parts of alcohol to 1 part of bisulphide of carbon are used for softening and reclaiming rubber. Dental and other gums are exposed to the sunlight in Alcohol to increase the brilliancy of the colors and to make the shades lighter. Alcohol is also used to soften vulcanized rubber when a surface color is to be added. Alcohol, in connection with nitric acid, spirits of turpentine, and aniline, was used by Kelly for surface work on India-rubber.

ANTHRACINE.—A trade name for napthaline (which see.)

BENZOL OR BENZOLE is a volatile oil obtained in the distillation of coal tar, which must not be confused with coal tar naphtha. Its specific gravity is 0.899 at 32° F., and 0.878 at 68° F. It is slightly soluble in water, and freely soluble in alcohol and ether, and in bisulphide of carbon. It is sold according to its percentage of pure benzol. It has great solvent properties. Benzol is used largely as a solvent for rubber in manufacturing bicycle cements, and also for dissolving rubber, and for the cold vulcanization of thin rubber fabrics containing chloride of sulphur, in which Benzol is much superior to carbon bisulphide; and at present it is much cheaper, both on account of less loss in handling, and also, of its much lower price per gallon. This refers more particularly to the high grades of Benzol, like 100 per cent. or C. P.; the 160° Benzol is mostly used where a solvent is required that must not evaporate too rapidly. It is said that if Gutta-percha is put in 20 times its weight of boiling Benzol, to which 1-10th of plaster is added, and the mixture agitated from time to time, a perfectly clear solution is decanted. This is then mixed with twice its volume of 90 per cent. alcohol and the Gutta-percha precipitated a pure white. (See Naphtha.)

BISULPHIDE OF CARBON is a transparent liquid, the specific gravity of which is 1.27. It is exceedingly volatile, evaporating at ordinary temperature. When properly made its smell is somewhat similar to chloroform. The bad smell found in some is due to sulphureted hydrogen, and the presence of foreign matters from which it can be thoroughly freed by purification. It is highly inflammable, though not explosive, and has great affinity for sulphur, 100 parts dissolving 37 parts of sulphur, cold; and at 100° F. the same quantity will dissolve 94.5 parts. Bisulphide of Carbon mixes with every known substance capable of vulcanizing rubber. It also assimilates rapidly with all fatty oils, and dissolves all the resins, with the exception of shellac. It does not dissolve vulcanized rubber, however. Where it is used in rubber factories care is taken, as a rule, to remove the fumes, as they are injurious to the workmen. Some very serious cases of chronic poisoning have occurred through the use of this solvent, the symptoms being numbness, partial paralysis, and, in some cases, temporary insanity. The use of Bisulphide of Carbon in rubber factories is very carefully watched, therefore, by the authorities in Europe, proper means for ventilation and carrying off the fumes being insisted upon, and minors being excluded from rooms where it is used. It is one of the best and most common solvents for India-rubber, very largely used in the Parkes cold curing and similar processes, and in cements.

Bisulphide of Carbon Substitute is a liquid produced by Dr. Carl Otto Weber, which is said to be a perfect substitute for bisulphide of carbon. It had these advantages: less chloride of sulphur was needed, the smell of the vulcanized product was sweeter, the vulcanizing solution penetrated deeper into the rubber, the risk of burning the rubber and the uneven vulcanization was also done away with. It is also said that this substitute is not injurious to the health. It is manufactured in England.

Borax is sometimes used as a solvent for rubber. (See Acids and Alkalies.)

Camphine is a name applied to one of the varieties of spirits of turpentine which was once largely used as a burning fluid. It is very volatile, and the vapor may exist in the air in explosive quantities. Camphine was formerly used to a certain extent as a solvent for India-rubber. Under Newton's method of recovering rubber, the waste was placed in a closed vessel, covered with Camphine, and heated to 158° F. or fourteen days. The solvent was then distilled off, and the tough mass remaining was capable of utilization, and was somewhat similar to unvulcanized rubber. It was also used in the boot heel cements in the old-fashioned method of attaching them to rubber boots, and also in general shoe cements. Camphine was also used in putting vulcanized waste, finely powdered, into a solution in connection with ether and alcohol, in a simple but somewhat expensive process of recovery.

Camphor has been used as a solvent for utilizing the

waste of vulcanized rubber and of hard rubber, the waste being first treated with any ordinary solvent and then placed in a still with a certain amount of camphor, when the India-rubber is dissolved and the solvent passed out and distilled over again. Granulated Camphor, over which had been passed sulphurous acid gas until it was reduced to a liquid, was used also as a solvent for India-rubber, by Alexander Parkes. (See Gums, etc.)

CAOUTCHOUCINE, also spelled Caoutchine, is a crude oil of India-rubber, made by its dry distillation, and smelling much like naphtha. It is an excellent solvent for India-rubber, but of course is too expensive for ordinary use. India-rubber immersed in it swells exceedingly, and a considerable quantity of it is dissolved during the boiling. It must be kept in hermetically sealed vessels, as it has a great affinity for oxygen, which it absorbs energetically. In preparing it, the India-rubber is treated in a retort at a heat exceeding 400° F. Caoutchoucine dissolves in ether or alcohol, and, absorbing oxygen freely, forms a resinous body as a result.

CHLORIDE OF CARBON.—This is obtained by the distilling of bisulphide of carbon into a vessel containing penta-chloride of antimony, the product being rectified by distilling with lime. According to Simpson, this makes a good solvent for India-rubber and in a measure vulcanizes it. Newton also used a chloride of carbon in dissolving both India-rubber and Gutta-percha, while Crump used tetra-chloride of carbon.

CHLOROFORM is prepared generally by distilling together a mixture of spirit—that is, wood alcohol—with bleaching powder, slaked lime, and water. Its density is from 1.496 to 1.498. It is one of the best rubber solvents known. It is costly, however, and has a bad effect upon workmen. Lascelles-Scott mentions what he calls the A. C. E. mixture which is composed of alcohol 15 parts, chloroform 38 parts, and ether 47 parts, which yields a powerful solvent for India-rubber or Gutta-percha. Chloroform dissolves not only India-rubber, but fats, resins, sulphur, alkaloids, and many other organic compounds. It should be remembered that a small percentage of chloroform in the air, even as little as 5 per cent., is dangerous to the workmen. Chloroform is used as the solvent for India-rubber which is treated with the ammoniac gas process for bleaching. Is also used alone, ~~and~~ in connection with naphtha for rubber cements, which are intended to adhere to glass. In the bleaching of Gutta-percha, it is also used as a solvent. One of the first uses of Chloroform in connection with India-rubber is to be noted under a patent granted to Charles F. Durant, who announced the discovery of a solvent known as "perchloride of formyle, otherwise known as chloroform."

CREOSOTE OILS, in connection with ordinary solvents for India-rubber, are said to produce a cheap and effective solvent. Indeed, John Bagnol, manufacturer for Charles Macintosh & Co., patented their use as applied to India-rubber. (See Creosote.)

DIPPEL'S OIL (or Bone Naphtha).—A thick, viscid oil of brown color and very disagreeable odor, which on distillation may be obtained limpid and colorless. It is prepared by the destructive distillation of bones, leaving boneblack as a residuum. It was one of the early solvents used for India-rubber.

ETHER.—This was one of the early solvents used in connection with India-rubber. It is sometimes called sulphuric ether, but erroneously. It is prepared usually by distilling a mixture of alcohol and sulphuric acid, washing the distillate, and rectifying the product with quick lime or something of that kind. It is a colorless, very mobile liquid, with a not unpleasant smell, burning taste, and very volatile. Its specific gravity is 0.736. It is soluble in water 1 to 12. Commercial Ether boils at 96° F., and yields a dense vapor.. It is very inflammable, and, when mixed with air or oxygen, gives rise to a dangerous explosive mixture. It is one of the best solvents known for oils and fats, and is also an excellent solvent for sulphur. For use in rubber work Ether should be free from water, but not absolutely pure, necessarily. It is little used to-day in rubber mills, except in some lines of very fine work. It has the advantage of being absolutely free from the smells that many solvents have. A little is sometimes added to ordinary rubber solutions to make a complete solution of India-rubber in naphtha. There are also certain processes, expensive ones to be sure, for treating perished rubber with Ether vapor to recover it. Ether was used to remove sulphur from vulcanized India-rubber waste in Newton's camphine process.

GASOLINE.—See Naphtha.

HEPTANE.—One of the four isomeric hydrocarbons of the paraffine series, which occurs as a colorless liquid and is derived from heavy cannel coal oil, petroleum, etc. Its specific gravity is 0.712. It is soluble in alcohol and in ether, and is used with paraffine wax and India-rubber in water-repellent compounds.

ISOPRENE.—A body which is found in oil of caoutchouc. It boils at 98.6° F., and possesses the property of absorbing quantities of oxygen when exposed to the air, in consequence of which it forms itself into an elastic spongy mass. This same volatile compound is obtained by the action of moderate heat on oil of turpentine. William A. Tilden, D. Sc., F. R. S., had some Isoprene from turpentine placed in a bottle, his first result being a limpid, colorless liquid. After a time, this changed in appearance, looking like a dense syrup, on which floated several hard elastic masses. On examination, they turned out to be practically India-rubber. This rubber united with sulphur in the same way as ordinary rubber, forming a tough, elastic compound. It was also soluble in benzine, etc. Dr. Weber, before the Society of Chemical Industry, reported on Tilden's discovery that Isoprene is so expensive it cannot be converted into rubber without loss, and therefore the synthetical manufacture of India-rubber, even if possible, was not probable at the present time.

LIGROIN.—See Naphtha.

METHANE.—Professor Lascelles-Scott describes the manufacture of what he calls Methane solvents, which are really benzines or benzols through which marsh gas has been passed. He claims that a benzine containing from 2 to 3 per cent. of Methane. obtained in this way, yields a better and more mobile solution than the ordinary solvent naphtha, and the solution when spread dries off better, besides giving a more finished surface.

METHYLATED ALCOHOL is also called methylated spirits, and wood spirits. It is obtained by the distillation of wood, and in the course of beet sugar manufacture. It is a colorless mobile liquid, of a vinous smell, similar to common alcohol. Its specific gravity is 0.814. It always contains acetone. Although not used in rubber solutions, it is a very common solvent for cellulose products which, through their increasing importance, are attracting the interested attention of the rubber trade. Used under Vaughn's patent, to coagulate Balata.

NAPHTHAS.—The term Naphtha was originally applied to a variety of pungent, volatile, inflammable liquids that belonged chiefly to a class of ethers; then it took in oils of natural origin, such as rock oil, petroleum oil, etc; at a later date, a light oil of coal tar, which should properly be designated benzol, was included under the name of Naphtha; while recently it has been extended so that it covers most of the inflammable liquids distilled dry from organic substances. It is applied in the United States to a series of hydrocarbons that are obtained from petroleum, whose boiling points vary with the densities, from 65 to 300° F. The Naphthas of commerce are bog-head naphtha, obtained from bog-head coal; bone naphtha, or Dippel's animal oil; coal naphtha, obtained from the distillation of coal tar; wood naphtha, or methyl alcohol obtained during the dry distillation of wood. Of these, coal tar naphtha and petroleum naphtha are most useful to rubber manufacturers. The former of these was used largely as a rubber solvent, but to-day it is almost wholly replaced by petroleum naphtha. The Naphtha which is derived from petroleum comes between gasoline, which is lighter, and benzine, which is heavier. Benzene is contained in the naphtha produced by the destructive distillation of coal, while benzine is a petroleum product. Benzine is really the first product that arises from the process of refining crude oil, and bears the same relation to naphtha that the distillate does to refined oil, thus showing that benzine is simply a crude Naphtha. What is known as gasoline has a proof rate of 86° F., and boils at 90° to 100° F. Warm currents of air volatilize this type of Naphtha very rapidly, and its vapor unites with the atmosphere in explosive proportions.

Coal-tar Naphtha was one of the first solvents used in rubber work. Macintosh, as far back as 1823, prepared it himself for dissolving India-rubber for proofing. There is obtained from crude Coal-tar Naphtha what is known as "once run" Naphtha and "last runnings." The once run Naptha is the starting point from which are derived the various grades of benzols, solvent Naphthas, etc., by fractional distillation. The specific gravity of solvent Naphtha should not exceed 0.875. Its composition is a very complex affair, including xylols, cumols, homologous of benzol, together with some paraffine, and sometimes a little naphthaline. This last-named substance, by the way, is often objectionable, as it acts upon some rubbers like animal oil. Naphtha derives its vegetable solvent power largely from the xylol present in it. This is to-day removed and sold by itself as a solvent, though the residual Naphtha is simply robbed of that much virtue.

Speaking of Naphthas, Lascelles-Scott, after exhaustive experiments, thus describes three used in England in rubber factories. Petroleum Naphtha in its solvent action on rubber showed slight action in the cold or under gentle heat. Viscid masses and semi-solutions were formed, but these solutions did not dry well. The same Naphtha had almost no solvent action on pitch. Shale Naphtha was useful only in dissolving Madagascar rubbers, and had no action on pitch, while coal-tar Naphtha caused almost any rubber to swell quickly and, after gentle heat, to effect a good solution. It also readily dissolved pitch, forming a deep brown solution.

The problem that confronts rubber manufacturers as a rule is the solution of gums that are more or less heavily compounded, which is an easier problem than the putting into solution of crude rubber that perhaps has not been broken down in any way. At the same time it is customary in many cases to apply a little heat during the mixing. The following table relates to petroleum Naphthas. The *C* Naphtha has not only the greatest solvent power, but it is easier to evaporate after it has dissolved the rubber compound. *B* and *A* require a certain amount of heat to vaporize them.

Products.	Specific Gravity.	Degrees Beaume.	Boiling Points.
Rhigolene	0.625	..	65° F.
Gasolene	0.665	85	120° F.
C. Naphtha	0.706	70	180° F.
B. Naphtha	0.724	67	220° F.
A. Naphtha	0.742	65	300° F.

Naphtha is more largely used in the proofing business than any other. It is, however, a general solvent for cements, and quantities of it are used in almost all lines of rubber work where there is any making up to be done of separate pieces after calendering. It is therefore necessary that a good grade be used, when one considers the danger that may come from fires caused by the explosion or easy ignition of low grade solvents. Odorless Naphthas are those from which naphthalene, a solid white body, has been removed, as it is the presence of this body that causes the strong smell. Naphtha treated by sulphuric acid is deodorized, acquiring a rather pleasant odor as a consequence. It is often mixed with other solvents—for example, with oil of turpentine—and is found thus to have a better effect on the rubber.

Naphthaline (called also Anthracine).—Commercially obtained from coal tar, being among the third and fourth products of the distillation of that body. Naphthaline is usually sold in rolls made by melting the large silvery plates or scales in which it crystalizes and running the melted compound into molds. Its specific gravity is 1.15. It is insoluble in water and petroleum naphtha, but the liquids derived from coal tar dissolve it easily. Naphthaline is sparingly soluble in alcohol and ether, but readily in benzol. It is used in insulating paints, as when it evaporates it leaves a very solid film that is said to be absolutely free from porosity.

Nitro Benzol.—A compound obtained by boiling benzol with nitric acid. It is a brown, heavy, oily looking liquid, having a specific gravity of 1.2, a burning sweet taste, and a smell resembling that of oil of bitter almonds. It is used in the analysis of vulcanized India-rubber to dissolve the substitute that may be incorporated in it. It is produced by the action of nitric acid on benzene, also called nitro-benzene. Used by Parkes in the manufacture of Parkensine. (See Acids and Alkalies; also Naphtha.)

Oil of Turpentine (crude) is what is known as an oleo resin, and is of about the consistency of fresh honey. There are more than a dozen varieties on the market, the more common being Bordeaux, Venice, Canadian, and American. A fair quality of turpentine oil should begin to boil at 160° F. The distillation of turpentine in water produces ordinary resin. Oil of Turpentine is used in certain waterproof cements, in connection with both Gutta-percha and India-rubber. Where oil of turpentine is necessary for rubber work, it is well to have it free from the considerable percentage of water which it invariably contains. This is done by a treatment with sulphuric acid, or by rectifying it over burnt lime. Turpentine, particularly that known as Venice Turpentine, is often used in connection with linseed oil and sulphur in the production of rubber substitutes. Professor Tilden showed, some years ago, that what appeared to be pure India-rubber could be obtained from turpentine; indeed, he announced that he had produced it on a small scale. The same thing was also observed by Bouchardt. Venice Turpentine is obtained from Switzerland, where it is procured from the *Larix Europea,* or larch. The genuine Venice Turpentine is of the consistency of honey, cloudy, yellowish, or slightly greenish. It is entirely soluble in alcohol. The commercial Venice Turpentine is a factious substance, usually quite brown, and is prepared by dissolving rosin in oil of turpentine. Venice Turpentine is largely used in cements. Bordeaux Turpentine is the ordinary turpentine of commerce, getting its name from the port in France whence it is exported. (See Spirits of Turpentine.)

PENTANE.—A hydrocarbon of the paraffine or methane series. A colorless, volatile liquid which occurs in petroleum. Pentane is used with paraffine wax and India-rubber in water-repellent compounds.

PETROLEUM.—A mixture of several hydrocarbons which, in fluid form, issue from the ground in many parts of the world; also known as rock oil. It varies in consistency from a thin, light, colorless fluid with a specific gravity of about 0.750, to a substance as thick as butter, and almost as heavy as water. All kinds, however, have about the same constitution, consisting of carbon and hydrogen compounds only, and containing no oxygen. Asphalt and bitumen are closely allied to petroleum. This oil is often used for restoring rubber that is oxidized somewhat, by immersion, and then hanging for a couple of days in a warm atmosphere. Petroleum is very rarely used in rubber manufacture, for although a good solvent, it weakens the goods exceedingly. Crude petroleum, however, is a valuable adjunct to the reclaiming of rubber, where, in the form of a cheap residuum, it assists in devulcanization and in sheeting. (See Naphtha.)

THION.—A substitute for bisulphide of carbon, manufactured in England, which is said to mix excellently with chloride of sulphur and is non-poisonous.

Toluene.—That oil which is distilled from coal tar at a temperature of 230° to 234° F., also called methyl benzine and Toluol. It resembles benzene in outward appearance. Two-thirds of the commercial 50 per cent. benzol is made up of Toluene, and this it is that makes it a far better solvent for rubber than benzine itself, as it dissolves the rubber in five-sixths of the time. The solutions are more mobile; it has a higher boiling point; and, given a quantity of the solvent, will reduce more gum. It does not chill in cold weather, but keeps on macerating. It leaves a more solid deposit than does benzine, and does not induce headache or sickness among the workmen. [Lascelles-Scott.]

Resin Oil.—This is obtained by subjecting rosin to dry distillation, the specific gravity of the resultant oil ranging from 0.96 to 0.99. It is rarely used as a solvent for rubber, in the ordinary meaning of the term. As a matter of fact, it is not a good solvent for crude rubber. For compounded rubbers, however, it also works well and is often used, particularly in connection with pseudo guttas. In certain insulating experiments, where a thin sheet of Gutta-percha covered the conductor, and the outer Gutta-percha tube was full of resin oil, it gave, according to Professor D. E. Hughes, F. R. S., a higher insulation test than Gutta-percha alone. Professor Hughes used resin oil quite thick and viscid, and added resin and a solid residuum obtained from the distillation of palm oil. Resin oil in rubber compounding, however, softens the compound in a marked degree. (See Oils.)

Rhigolene.—See Naphtha.

Spirits of Turpentine is really oil of turpentine, and it has a specific gravity of 0.864. It is colorless, transparent, of a strong odor, and a bitter taste. It is insoluble in water, on which it floats, but readily soluble in alcohol, ether, and the fixed and essential oils. It is an excellent solvent for sulphur, resin, and India-rubber. Spirits of turpentine, with wood spirit alcohol, aniline, and nitric acid is used in surface work on vulcanized India-rubber. The earliest records of India-rubber speak of this oil as a solvent for it; indeed, the whole secret of rubber compounding for a number of years, even when the great Roxbury Rubber Co., of Boston, was running, was the solution of India-rubber in it. It is used in solutions that are expected to be sticky, and to dry slowly.

VULCOLEINE is a liquid of English origin, and is put upon the market at about the same price as carbon bisulphide, and used for a solvent for India-rubber. It leaves on evaporation a perfectly tough and elastic film, quite unlike that left by coal tar naphtha, or the usual solvents. It mixes instantly with chloride of sulphur, and is intended to replace bisulphide of carbon in the cold curing process. It has no bad smell, nor is it unhealthful.

WOOD SPIRIT (also known as Pyroxylic acid).—This is made from the destructive distillation of wood. Wood Spirit resembles alcohol and its affinities, forming an ether and a series of compounds exactly corresponding to that of spirits of wine. Wood Spirit, when pure, is a thin, colorless liquid, with a peculiar odor and a hot disagreeable taste. It boils at 152° F., and its density is .798 at 60°. It mixes freely with water, and, like alcohol, dissolves resins and volatile oils, and is used as a cheap substitute for that purpose. Wood Spirit, also known as methylic alcohol, is not methylated spirit. It is not a solvent of rubber, but is used in many compounds that are intended as substitutes for vulcanized rubber. It is also used in dyeing India-rubber in connection with nitric acid, alcohol, and aniline.

XYLOL.—A colorless, somewhat aromatic, inflammable, oily liquid found in coal tar and wood tar; also called Xylene. It is really the solvent principle found in mineral naphthas. (See Naphtha.)

CHAPTER XII.

MISCELLANEOUS PROCESSES AND COMPOUNDS FOR USE IN THE RUBBER FACTORY.

MANY interesting formulas are given for the dyeing and surface coloring of rubber, although the processes are not such as will generally be used. A suggestion that comes from France is the dipping of rubber for an instant in a bath of nitric acid, then washing in water. For coloring, the rubber is dipped in an alcoholic solution of fuchsine. The experimenter should appreciate fully, however, the effect that nitric acid produces on rubber, and govern himself accordingly.

Alexander Parkes, who produced some exceedingly valuable processes for the treatment of rubber, gives the following formulas for dyeing India-rubber:

Black.—Boil from 15 to 30 minutes in a liquid prepared as follows: Sulphate copper, 1 pound; water, 1 gallon; caustic ammonia or muriate of ammonia, 1 pound. Or: Sulphate of bisulphate potash, 1 pound; sulphate copper, 12 pounds; water, 1 gallon.

Green.—Muriate ammonia, 2 pounds; sulphate copper, 1 pound; caustic lime, 4 pounds; water, 1 gallon. Boil the rubber as before, 15 to 30 minutes.

Purple.—Sulphate or bisulphate of potash, 1 pound; sulphate of copper, ¼ pound; sulphate of indigo, ¼ pound. Boil the rubber, 15 to 30 minutes.

Hoffer gives almost the same ingredients for producing these colors, adding the information that the articles are dyed by being boiled in these fluids from 15 to 30 minutes, the thicker the article the longer the boiling. This is done before the goods are vulcanized.

Hard rubber may be decorated by means of pigments mixed with shellac and applied to the given surface with a brush. The surface then is to be pressed with some force against a hot plate of metal, whereby the colors are made to appear as though integral with the rubber.

Wood coated a sheet of vulcanizable rubber with chloride of

silver, the idea being to use it in dental plates. Various processes have also been brought out for the surface treatment of rubber with gold leaf, bronzes, etc., usually applied in the form of powders, in the manner in which flock is applied. Truman also patented a process for electro-gilding rubber dental plates after they were finished. Goodyear dusted unvulcanized rubber surfaces with plumbago or powdered metal, to make them conductive, pressed the dust in, and then electroplated it.

The embossing of India-rubber surfaces has been practised almost since the invention of the "triple compound." It is really nothing more than a light surface molding. This is done sometimes by embossing rolls, the rubber being cured after the impression is taken, and sometimes by being vulcanized on the impression plate.

Bourbridge patented a process for embossing rubber by rolling it tightly on a drum with embossed paper or bookbinders' cloth, and semi-curing it in that form, preferably by boiling at a temperature from 212° to 220° F. This boiling operation was not really vulcanization, but simply a means of setting the rubber which was afterward made up into goods and cured.

In producing sheets of India-rubber for the manufacture of tobacco pouches, balls, balloons, etc., by this process, the sheet is calendered on sized cloth, partially vulcanized, printed, coated with transparent India-rubber, the goods made up, and the vulcanizing process completed.

A great many beautiful colors are added to India-rubber surfaces by coating the sheet with a thin adhesive solution, dusting it over with colored flock, and then vulcanizing. By this process any color can be given to rubber surfaces which have a cloth-like appearance.

Kelley produced a bronzed appearance on rubber coated fabrics by means of a roller partly immersed in a trough holding the dye, curing either by dry heat, or by chloride of sulphur. His solution consisted of 2 ounces alcohol spirits, 1 ounce wood naphtha, 10 drops nitric acid, 1 ounce spirits of turpentine, with sufficient aniline dye to make the desired color, 4 ounces liquid dyeing, 3 pounds rubber composition. He also impregnated farina with aniline solutions, dried it, and mixed it in the compound.

In certain dyeing processes lakes are necessary. What is known as caoutchouc lake is made by steeping 1 ounce of Para rubber in a quart of light camphor oil, exposed to the sunlight for several days. This is said to be excellent for binding colors.

Matthew's process for producing colored designs for proofed fabrics is to first coat the fabric in the ordinary manner with pure or colored India-rubber. When the design is to be printed on a black or dark ground, the last coating is mixed with starch or some powder that will render it non-adhesive, and to an extent absorptive. The fabric is then partially vulcanized, when the designs are printed on the desired surface, just as oil-cloth or linoleum is printed. The vulcanization is finished preferably by using chloride of sulphur.

Colors suitable for admixture with rubber should answer the following requirements: They must be unaffected by water, by acids, by alkalies, and by chloride of sulphur. Further than this, they must not be affected by sulphur at temperatures ranging from 200° to 300° F. The colors must not be soluble in or affected by naphtha or other solvents used in rubber work. They must not be affected by heat up to 300° F. According to Frankenburg, his invention of aniline lakes answers all these requirements. His description is as follows:

(A) Lakes prepared from acid aniline colors.—"I have found that by converting any of the acids or sulphonated aniline colors into compound lakes, such as barium-alumina, calcium-alumina, barium-chromium, or calcium-chromium lakes, colors are obtained answering all the above requirements, and therefore eminently suitable for the dyeing of India-rubber, waterproof, and other articles. The aniline dyes best suited for the production of these lakes are those known as azo or dis-azo colors. From colors of this description I prepare lakes in the following manner: 50 pounds of orange II., or any other suitable azo or dis-azo color, and 112 pounds of soda crystals are dissolved in 100 gallons of water at 170° F. This solution is then precipitated with a solution of 150 pounds of barium chloride. The precipitate is kept boiling for half an hour. It is then left to stand, and washed several times with fresh water. Eventually a solution of 40 pounds of alumina sulphate is added very gradually, when a bright, fast, and flocculent lake is obtained, which, after filtration, drying, and pulverizing, is ready for incorporation with the India-rubber dough. It is evident that a great many variations of the process may be devised, but in every case the important point is the conversion of the aniline dye into one of the above-mentioned compound lakes. As regards the proportions given above, they are, of course, subject to such variations as are in accordance with the molecular weights and the commercial purity of the materials used, as well as with the particular properties and qualities to be imparted to the lakes for the purpose they are intended to serve. Using in this manner the numerous azo and dis-azo dyes a very great variety of lakes may be produced, comprising all conceivable shades, and all suitable for the dyeing of India-rubber articles of every description. The lakes prepared from the acid oxy-ketone dyes and most of the natural dyes are very little suitable for this purpose, owing to their indifferent and dull shades."

(B) Lakes prepared from basic coloring matters.—"A large number of lakes derived from this class of dyes are also suited for the dyeing of India-rubber articles, although many of them are lacking in fastness to light acids and alkalies. To produce a perfect compound lake from these dyes tannic acid and antimony, along with aluminum and barium, are used for the complete fixation and precipitation of these lakes. The following proportions give good results: Soda carbonate, 128 pounds; barium chloride, 110 pounds; thioflavine, 25 pounds; tannic acid, 20 pounds, acetate of soda, 20 pounds; sulphate of alumina, 100 pounds. These colors can be made faster by adding to them a small quantity of antimony potassio-tartrate. The proportions of tannic acid, sodium acetate, and tartar emetic used in this process vary considerably with the different basic colors, such variations being due to the difference in the atomic weights and commercial purity of the basic dyes."

Hebblewaite and Holts's process for producing designs on gossamer cloth calls for the spreading over the rubber surface of farina or other powder, then running the fabric through embossed rollers and producing patterns thereon.

Mosley's ornamented fabric was a gossamer cloth covered with farina, the surface being printed much as calico is, and then vulcanized with chloride of sulphur. The colors were mixed with suitable solvents and a certain amount of paraffine or India-rubber added. A part of this invention was also the use of an engraved roller, which revolved in the vulcanizing solution, and came in contact with the surface of the rubber, only at its raised portion. Directly after passing over the roller, if the surface of the rubber were dusted with farina, it would adhere to the portions that had come in contact with the roller, and not to the rest, thus producing a design on the fabric. The whole of the coating was afterwards cured by vapor.

SHOWER-PROOF PROCESSES.

THE Cravenette and other processes for rendering textile fabrics waterproof or water-repellent have attracted so much attention in the rubber trade that space will be given here to a description of the Wiley patent, which is used at the Cravenette Works, Bradford, England. To begin, the waterproofing compound is applied in a solid or hard state by the action of friction and heating. In other words, there are no solvents used, nor is it a calendering process. The advantage of this is a lessening in the cost of applying waterproofing solutions and a further valuable result is that the dyes on various fabrics are in no way disturbed, and no unpleasant odor is developed or imparted to the cloth. The substances chosen are those which have a low melting point, so that the fabrics are not damaged by heat. They are preferably ozocerite, stearine, spermaceti, paraffine wax, beeswax, or Japanese wax. These are sometimes used singly, and sometimes in combination, considerable judgment being necessary in selecting those which have an affinity for or are readily absorbed by the fibers of particular fabrics, influenced also by the nature and color of the fabric. In some cases India-rubber, Gutta-percha, maltha, asphaltum, resin, and artificial gums are found valuable in small proportions, and in conjunction with the substances already mentioned.

In order to apply the waterproofing substance, it is formed into slabs. The fabric is carried on a reel supported in bearings between suitable frames, at the opposite end of which is a hollow cylinder mounted upon carrying rollers and supported laterally by side rollers. This cylinder is filled with water. The slab of the compound, wider than the fabric to be coated, is fixed in a holder above the cylinder. This holder is so arranged that the weight presses the slab against the cylinder. The fabric is then drawn from the reel over and under tension bars, under a supporting roller, between it and the rubber cylinder, and around the cylinder and under the slab, then over the guide roller and into a drying machine. The friction of the cloth wears the slab away and uniformly deposits it upon the cloth, while in the drying machine, the heat melts the waterproofing compound, and it is absorbed by the fibers which are thereby rendered waterproof or water-repellent.

Other formulas for shower-proofing and waterproofing are of interest in this connection and a few are given:

The first is a German waterproofing compound: Alum, 10 pounds; sugar of lead, 10 pounds. Dissolve in hot water and allow the precipitate to settle. Dilute the clear liquid with 120 gallons water and add 2 pounds isinglass in solution. The goods are steeped in this solution 8 or 10 hours.

An American shower-proof compound: Liquid silicate of soda or liquor of flint, 1 gallon; white oxide of zinc, 1 pound. If the fabric is to be colored, add coloring matters. The mixture may be applied to fabrics hot or cold, by means of a brush or by immersion of the fabrics, which are afterwards to be run between rollers.

Another American compound: Dissolve separately, 1¼ pounds alum (in hot water), 10 ounces acetate of lead (in hot water), and 1½ pounds carbonate of magnesia (in hot water). They should aggregate about 31 quarts. Add the acetate of lead to the alum solution, and then the carbonate of magnesia; after which 10 quarts liquid as above and 1 tablespoon white gum arabic. Stir ½ hour; let stand 24 hours, skimming now and then; in 48 hours the first mixture will be ready. Lay the fabric in a vessel and pour liquid over it, beating the fabric well and removing it within an hour.

A third American shower-proof compound:

A.	Carbonate of soda	16 parts.
	Lime	8 parts.
	Water	32 parts.

Boil 30 minutes, let settle and pour off the clear lye.

B.	Glue or gelatine	3 parts.
	Linseed oil	3 parts.

Add after soaking glue in cold water 12 hours.

C.	Tallow (or other animal fat)	16 parts.
	Rosin	8 parts.

Melt together.

To (A) boiling hot add hot (C), then pour in (B) and stir hot until well mixed.

D.	Sulphate of alumina	1 pound.
	Acetete of lead	½ pound.
	Boiling water	8 gallons.

Let settle and draw off clear liquor for use. To 1 gallon water add ½ ounce of first product for bath for cotton goods. Add ¼ ounce for silk or wool. Immerse 24 hours or more, then six hours or more in second compound (D).

Proofing compound:

Mixture 1.—Dissolve in water, 50 parts alum; also dissolve in water, 35 parts sugar of lead; mix.

Mixture 2.—Combine 17 parts paraffine and 35 parts benzine; drop into this 17 parts Caoutchouc. Stir until well dissolved.

Mixture 3.—To the clear decanted liquor from the above mixtures, add 8 parts alcohol and 4 parts eau de cologne (or oil of lemon.)

An English compound for waterproofing textile fabrics: Sugar soap, 1 pound; water, 16 gallons. Soak articles in them for 6 hours; drain, but do not wring them; and place them in the following solution:

Alum, 1 pound; water, 16 gallons; soak again 6 hours, take out and dry without wringing.

Another English compound for waterproofing textile fabrics: Concentrated size, 8 pounds; aluminum sulphate, 5 pounds; barium chloride, 6 pounds; water, 16 gallons. After coating, varnish with the following: Melt together 22 pounds colophony, 4 3-5 pounds crystalized soda, and 11 pounds water. Then add: Ammonical fluid, 5¼ pounds; and water, 55 pounds; or: Borax, 6 pounds; shellac, 6 pounds; and water, 40 pounds.

A German compound for waterproofing woolens: Dissolve 100 pounds alum in moderate quantity of boiling water; soak 100 pounds glue till it has taken up twice its weight of cold water, then apply heat to dissolve it; stir 5 pounds tannin and 2 pounds soluble glass well into the glue, then add the alum solution. Enter the goods at 80° C., and steep 30 minutes. Take out and drain several hours, stretch on a frame, and, when dry, calender.

A German shower-proof compound: Stir 9 pounds casein well in 32 quarts water, adding little by little 25 pounds of slaked lime. Add a solution of 4½ pounds soap in 26 quarts water.

Filter and treat the cloth with the liquid. Dress with a dressing of acetate of alumina, by which the casein is rendered insoluble in the fibers of the cloth. After two applications, rinse the goods with hot water, press strongly, and dry.

One process for waterproofing threads and yarns used in weaving ducks and other fabrics is in two parts, the first of which relates to a tanning mixture in which the yarns are immersed, consisting of: Birch bark, 14 pounds; bichromate of potash, 1 pound; chloride of calcium, ½ pound; tar 1 pint; solution of alkali, 2 pounds. The threads are first boiled in a 5 per cent. solution of alkali to destroy perishable matter, after which they are immersed in the tanning liquid and dried. The second part consists of preparing or dressing the threads with the following compound:

Poppyseed oil, 2 gallons; India-rubber solution, 2 pounds; red oxide of mercury, 1 pound; resin, 28 pounds; beeswax, 28 pounds; palm oil, 14 pounds. The threads after this treatment are wound on reels for weaving.

Forster, as far back as 1847, made a water-repellent compound in which he used spermaceti, wax, and stearine, while three years prior to that Townsend used two solutions to accomplish that end, the first being water, calcined British gum, white soap, logwood liquor, and rock alum; the second being water, sulphate of zinc, calcined British gum, and palm soap.

The Kyanized cloth process is well known in connection with preserving fabrics, the treatment being with a mixture of corrosive sublimate, chloride of zinc, pyrolignite of iron, oil of tar, and resinous matters. Fabrics treated in this way have been used for the manufacture of hose,.

Crape cloth is a fabric which has much the appearance of real crape, but is far less expensive. It is treated with processes similar to the Cravenette process, which make it both waterproof and durable. Two patents for this process have been granted to W. E. B. Priestly.

According to Dr. Doremus the lightest fabrics are rendered uninflammable by dipping them in a solution of phosphate of alumina in water.

Allard's fireproof felt is made of 50 per cent. of asbestos and 50 per cent. of animal hair, and for ordinary purposes is wholly fireproof.

Canvas for sails and other purposes, which it is desired to render waterproof, is treated by the Dumas process so that, while it is both waterproof and fireproof, it is still elastic and permeable by air. The treatment is this: The material is first put in a solution of gelatine, then run through pressure rollers ,and spread in the open air to dry; later it is dipped in a cold solution of alum again exposed to the air, then washed in cold water, and finally dried.

Frankenburg's waterproof cloth is made in this manner: Both warp and woof are coated in the yarn with India-rubber, then powdered with farina, then woven, after which the fabric is calendered, and the result is a cloth that is thoroughly waterproof, and yet does not give evidence of having rubber in its make-up.

Smith's porous waterproof fabric called for a compound made of 100 parts of paraffine melted by heat, to which was added 15 per cent. of India-rubber, the mixture being kept from 5 to 30 minutes at a temperature of 100° C. The solution, either as it is, or with a solvent, is then transferred to the cloth by means of a set of rollers which have a temperature of about 70° C.

DEODORIZATION.

THE odors that cling to vulcanized rubber goods and to Gutta-percha are often very objectionable, and the following processes are given for deodorization:

Cattell's process: For every pound of well cleaned Gutta-percha take 15 pounds of the following solution: Benzole, 1 gallon; alcohol, 1 ounce; glycerine, 30 drops. Or: Benzole, 1 gallon; nitrate of the oxide of ethyl, 30 drops; heat in a closed vessel to 110° F. The Gutta-percha is recovered by cooling to below

32° F., and pressing or by distilling off the solvent, or by precipitation with fusel oil.

Freeley's process: Dip vulcanized rubber goods in a solution of: Salicylic acid, 20 grains; alcohol, ½ pint. This will deodorize them, but goods will be toughened and the deodorization increased by subjecting goods to a bath in hot or cold solution composed as follows:

(A) Bark of oak, 50 pounds; bark of hemlock, 50 pounds; bark of sumac, 50 pounds; water, 900 gallons.

(B) Solution as above, 2 gallons; salicylic acid, 20 grains; large tablespoonful of Russian Jackten extract, dissolved in 2 pints of alcohol, 1 pint of ether, and 10 grains of salicylic acid.

Bourne's process: The articles to be deodorized are placed between layers of charcoal and heated from 120° to 150° F., if unvulcanized; 180° F. if partially vulcanized; or 212° F., if completely vulcanized. Heat for six hours or more.

Lavater and Tranter's process: Subject the articles to a boiling in potash, then to a vacuum, then to a pressure of air scented with some essence. They claim the extraction of the sulphur from the pores of the rubber in the form of sulphuretted hydrogen and its replacement by perfumed air.

Charles Hancock's process: To remove the odor of Gutta-percha, steep it in the following solutions:

(A) Soda or potash, 1 pound; water, 10 gallons.

(B) Chloride lime, 1 pound; water, 10 gallons.

De la Granja's process:

Iodine	15 grains.
Permanganate of potassa	20 grains.
Iodide of potassium	60 grains.
Glycerine	4 ounces.
Sulphite of soda	4 ounces.
Sulphite of lime	4 ounces.
Sulphite of potassa	4 ounces.
Water	1½ to 2 gallons.

Steep or macerate rubber in a solution composed as above, in a close earthern vessel, 24 hours, the solution being cold. Then heat the solution gradually to boiling point and uncover the vessel until ⅛ of weight of solution evaporates. When the solution cools remove the rubber.

PRESERVING RUBBER GOODS.

THE deterioration of vulcanized rubber goods is often a serious matter, where it is necessary for some time to keep them in store. Wherever possible, they should be kept in a cool dark place, and away from warm currents of dry air. It has been advised that such goods as druggists' sundrics be stored in an air-tight receptacle, in the bottom of which is placed a vessel containing benzine, which is allowed to evaporate slowly. Kreusler and Bude in *Der Techniker* recommend the dipping of the articles in a paraffine bath, heated to about 212° F. This does not injure the color or the appearance, but is said to enable the goods to effectually resist both light and atmospheric influences. From its well known softening effect on India-rubber, however, paraffine is likely to be used with considerable care by rubber manufacturers. In the line of mechanical goods, Turner patented a process for treating both hose and tubing with carbolic acid, either during its manufacture or after vulcanization in order to preserve it. Torrey also saturated duck with carbolic acid before it was made up into hose.

Mowbray's process for preserving rubber in valves: The use of 20 pounds of India-rubber, washed and cut fine, in connection with 5 to 10 pounds of naphthaline; digest 24 to 48 hours, at 180° to 230° F. Masticate in a machine heated to 212° F., until it forms a plastic homogeneous compound. If other substances are to be added, treat as follows:

1. Soluble matters (sulphur, antimony, resins, etc.) dissolve in naphthaline, melted or boiling, and add to above naphthalized caoutchouc at temperature of 240° F.

2. Materials insoluble in naphthaline (oxides of lead and zinc, chalk, etc.) deprive of moisture and heat to 212° F. and add to naphthalized caoutchouc.

This compound can be used for soft or hard rubber, according to the proportion of sulphur used. The object is to preserve the elasticity of rubber and prolong its durability.

Trueman's process for preserving India-rubber, and fibers that may be used with it, employs the peroxides of manganese and lead and the black oxide of copper, all of which have the property of decomposing ozone in great quantity, and converting it into oxygen. The inventor believes that ozone is the active agent in producing decay, and, by changing it into oxygen, he arrests such decay. In applying these oxides, he mixes them with ozocerite or tar.

Elworthy patented a process for storing rubber goods in a receptacle filled with nitrogen, hydrogen, marsh gas, or carbonic acid gas. This was recommended especially for rubber goods in India.

FASTENING RUBBER TO METALS.

THE problem often comes to rubber manufacturers as to how to stick rubber or rubber compounds to iron so that they will not part from it, no matter under what strain. This is done successfully by a number of different formulas. Where the processes are skilfully carried out, the rubber should adhere so firmly to the iron, that it will disintegrate and give way anywhere else in the mass, except where its surface is in contact with the metal. The basis of all these processes is said to be the chemical affinity for sulphur which is in the rubber with the copper salts used in the compound. One formula for this is: First, the grinding of the iron, finishing it with a file, and dipping it in strong lye to remove all grease, and afterward in muriatic acid or dilute sulphuric acid heated in water. The metal is cemented before the rubber is applied.

The process patented by Garrity and Avery, is as follows: Nitric acid (41° Baume), 10 gallons; muriatic acid (22° Baume), 10 gallons; mix and add pure tin, finely divided, 10 pounds.

Immerse the iron for 8 seconds, remove and dip into weak solution sulphuric acid, then wipe with a woolen cloth. Then apply with brush or otherwise, the following compound: Rubber cement, 7½ gallons; litharge, 6 pounds; and sulphur, 3 pounds. Add vulcanizable rubber compound at once, and vulcanize.

Hall's process: Water, 100 quarts; caustic potash, 10 pounds; cyanide of potash, 2 pounds; sulphate of copper, 2 pounds; sulphate of zinc, 2 pounds. The pickle and bath are made of water and about 10 per cent. sulphuric acid, the tub being lined with brass plate.

Adams's process: A weak solution of sulphate of copper is made—say 2 or 3 ounces of the crystalized salt to the gallon—and this solution may be acidulated with sulphuric acid—say about ½ gill of strong acid to the gallon. For a fine film for "dipping" articles of iron, steel, or tin, to which the rubber compound is to be applied, if the metal is copper, it should first be coated with tin, nickel, or iron.

The Shellac process calls for a cement made of shellac steeped in ten times its weight of concentrated ammonia, the solution being allowed to stand three or four weeks. This solution is painted on the iron, allowed to dry, and the rubber vulcanized upon it.

THE USE OF GASES.

Before India-rubber reached its present value in the arts, and before coal gas was generally known as an illuminant, Mollerat obtained oil of caoutchouc by distillation and made a fine quality of illuminating gas from it. It is needless to say that the process is not practised to-day.

Pellen rendered India-rubber impervious to gas by coating it with collodion mixed with a very small quantity of castor oil or with a varnish composed (1) of 32 per cent. of gum arabic, 8 per cent. of sugar, and 60 per cent. of water, or (2) made from 28 per cent. of dextrin, 60 per cent. of water, and 12 per cent. of gelatine.

Bousfield rendered vulcanized India-rubber impermeable to gas by applying linseed oil to it in the form of a varnish, the articles being heated.

Parkes suspended articles to be vulcanized in a dry heater and passed the following gases into the chamber as a means of vulcanization: Sulphurous acid gas, chlorine, nitrous acid, or the vapors of bromine or iodine.

Charles Hancock cured rubber by the action of vapors produced by dissolving zinc, copper, or mercury in nitric acid. The action of these vapors being so solvent, only one or two moments were given, and the surfaces then washed in an alkaline solution.

Nickels passed sulphur fumes and hydrogen into the gum while in a masticator, curing afterward by heat.

Johnson prepared carburet of hydrogen from oil of tar as a solvent for Gutta-percha. In order to overcome the smell of the solvent, he added a little alcohol in which was essence of lavender.

Hughes made an artificial rubber from gelatine, resin, oil, and tannin, improving the compound by exposing the compound to the action of hydrogen, sulphurous gas, sulphuretted hydrogen, nitrous gas, or ammonia.

Brooman treated vulcanized waste rubber with vapors of turpentine in his reclaiming process.

Lake bleached India-rubber in a stream of ammonia gas or chloride of ammonia, afterwards thoroughly washing the gum in hot water.

A great many rubber goods—that is, thin sheet goods—are cured by what is known as the vapor process. This is done in many cases by hanging the goods in an air-tight chamber, like a dry heater, and passing the vapor, which is either that of chloride of sulphur alone, or chloride of sulphur mixed with nitric acid, into the curing room. Small articles are often put in a tumbling barrel made of wire, which revolves slowly in the vulcanizing room, thus giving the vapor a chance to do its work thoroughly. The rubber surfaces are of course dusted first, to keep them from adhering. Proofed cloth is cured in vapor by passing the rubber surface over troughs in which this reagent is slowly evaporating.

The vapors of ozocerite are also used in rendering cloth water-repellent.

A mixture of chlorine and hydrogen gas is used for filling small India-rubber balloons. A fuse is attached to which a spark is applied before it is let off. After a time this spark reaches the gas, and the balloon explodes.

Vulcanized India-rubber, whether compounded or pure, is permeable by gas. In making flexible gas tubing, therefore, it must be coated or in some way protected in order to make it gas tight. The common way of accomplishing this is to cover the rubber tube with an outer tube made of glue, glycerine, and bichromate of potash, this covering being protected in turn by a woven fabric. Another plan for accomplishing the same result is to have an outer and inner tube of India-rubber, between the two being vulcanized a sheet of tin-foil.

ACTION OF METALS ON RUBBER.

THE action of various metals on India-rubber has always interested rubber manufacturers. In the memoirs and proceedings of the Manchester (England) Literary and Philosophical Society, 1890-91, William Thomson, F. R. S., and Frederick Lewis published an exceedingly interesting paper on this subject. They covered almost all of the metals that are likely in any way to come in contact with rubber surfaces, and proved what has long been acknowledged by rubber manufacturers, that the action of copper is most harmful. The metals that have no action at all on rubber are gold, silver, bismuth, antimony, arsenic, tin, chromium, iron, nickel, cobalt, zinc, and cadmium. Those that act only in a slight degree on rubber are lead, aluminum, palladium, and platinum.

Of the salts of metals that are very destructive, copper stands first, manganese oxides and nitrate of silver, being, however, almost as bad. Several other nitrates have also an injurious effect, although not as much so as those just mentioned. They are the nitrates of ammonia, uranium, sodium, and iron.

According to N. Foden, a well-known English expert, proofed goods in browns have caused him more trouble by deterioration than any other colors—more than black, even—and it is to be said right here that blacks as a rule are viewed with distrust by manufacturers, because it is believed generally that copper salts are used in the dyeing. Mr. Foden instances the time when brown tweeds were used largely, and when most manufacturers experienced a great deal of trouble with them, as the browns showed early signs of decay, while the grays remained soft and flexible. Mr. Foden suggests that, as certain dyers use lime, which is cheaper than logwood, this may act destructively upon the rubber.

ARTIFICIAL RUBBER MILK.

WHEN rubber in solution of almost any of the ordinary solvents is mixed with a moderately large quantity of methylated spirit, it is precipitated and forms later a sticky, whitish mass from which the resins and coloring matter have been taken by the spirit. Instead of this process, Lascelles-Scott advises the following: Take a 10 or 15 per cent. solution of fine Para rubber in benzine or chloroform with a little strong alcohol, but not enough to precipitate the rubber. If a considerable volume of tepid water be then quickly stirred into the solution, the rubber slowly separates from its solvent. If to this is added a little resin-potassa soap, with a little liquor ammonia, the emulsion is very similar to rubber milk. The distinguished author suggests the use of potassa soap made of the native rubber resin as the best emulsifying compound for such a purpose.

In writing on the preservation of genuine rubber milk, he also condemns the use of creosote, for, although it prevents fermentation, it does not hinder the gum from separating. He advises the use of ammonia and if it is to be kept through hot weather, the addition of a fragment of camphor or naphthaline or a few drops of santal-wood oil.

SHRINKAGE OF RUBBER.

THE following table shows the average rate of shrinkage in the various leading grades of India-rubber, and also the widest range of shrinkage noted in the practice of some extensive manufacturers. The figures express percentages in weight:

	Average.	Range.
Para sorts:		
Fine	16 to 18	15 to 20
Medium	17 to 19	16 to 22
Coarse	22 to 28	18 to 35
Mangabeira	25 to 30	20 to 35
Caucho	26 to 34	20 to 40
Centrals	26 to 32	20 to 40
Africans:		
Tongues	19 to 24	18 to 25
Flakes	28 to 33	25 to 35
Thimbles	22 to 28	15 to 35
Accra sorts	24 to 32	20 to 40
Congo sorts	19 to 24	18 to 35
Benguella sorts	16 to 20	16 to 20
Mozambique sorts	17 to 28	10 to 35
Madagascar sorts	30 to 40	25 to 55
Assam	23 to 31	8 to 45
Borneo	33 to 38	30 to 45

Mr. T. Bolas, in his "Cantor lectures" on India-rubber, in 1880, gave the following estimates of shrinkage of these leading grades:

Para	15 per cent.
Para negroheads	25 "
Ceara	28 "
Guayaquil	40 "
Borneo	25 "
African ball	25 "
African tongues	35 "
African niggers	25 "
Madagascar	25 "

PARA RUBBERS.

The next table indicates in detail the percentage of shrinkages in the various grades of Para rubber, also determined by the practice of American manufacturers:

	Fine.	Medium.	Coarse.
Bolivian	15 to 17	16 to 18	20 to 25
Mollendo	15 to 17	16 to 18	....
Madeira	15 to 18	16 to 19	20 to 25
Manaos	16 to 17	17 to 18	18 to 22

Upriver	16 to 18	17 to 19	18 to 25
Matto Grosso	16 to 18	17 to 19	20 to 28
Angostura	16 to 18	17 to 19	25 to 30
Caviana	16 to 18	18 to 20	25 to 30
Itaituba	17 to 18	18 to 19	20 to 25
Islands	18 to 20	18 to 22	25 to 35
Cameta	....	....	30 to 35

The shrinkage of Mangabeira (Pernambuco) thin sheet is about 25 to 30 per cent.; thick sheet, 30 to 35; ball, 20 to 25. Caucho (Peruvian) slab, 30 to 40; sheet, 30 to 35; strip, 25 to 35; ball, 20 to 25.

The better grades of Centrals shrink from 25 to 30 per cent.; other grades, generally from 30 to 40.

AFRICANS.

The Gold Coast sorts (including Accra, Cape Coast, Saltpond, Addah, Quittah, and Axim) range about as follows: Buttons or biscuit, 20 to 30; flake, 30 to 35; lump, 30 to 40; niggers, 20 to 35.

Cameroon ball, 18 to 25; clusters, 18 to 28.

Lagos buttons, 25 to 35; lump, 30 to 40; strip, 25 to 35.

Congo buttons, 25 to 30; ball No. 1, 20 to 25; ball No. 2, 25 to 35; Upper Congo ball and strips, 20 to 25; red ball, 18 to 22; Equateur small ball, 16 to 20; mixed ball, 18 to 22; Lopori small ball, 16 to 22; Kassai black twist, 18 to 22; red twist, 20 to 25; ball, 20 to 25.

Benguella (and Loanda) sausage, 16 to 20; niggers, 18 to 20.

Mozambique (including Lamu) ball No. 1, 10 to 15; ball No. 2, 15 to 25; ball No. 3, 25 to 35; sausage, 20 to 35.

Madagascar pinky, 30 to 35; Majunga, 30 to 35; black, 30 to 40; niggers, 30 to 40.

EAST INDIAN.

Assam No. 1, 10 to 15; No. 2, 20 to 30; No. 3, 30 to 35.

Penang, No. 1, and Java No. 1, 10 to 15 per cent.; other numbers same shrinkage as Assam.

E. Chapel gives this table of percentages of shrinkage:

Para, fine	12	Ceara	28
Para, coarse	25	African ball	28
Loando	17	Madagascar	28
Colombia	20	Assam	28
Java	22	Gaboon	35
Gambia	24	Borneo	35

TO FIGURE SHRINKAGE IN CRUDE RUBBER.

It is strange that there should be a divergence of opinion and method in arriving at the net cost of rubber after washing, sheeting, and drying it, yet such is the case. To assist those who have not studied this question, the right and the wrong way of figuring on shrinkage is given here. Take for instance an average-priced rubber:

Example A.

100 lbs. rubber at \$0.50 = \$50.00
20 lbs. shrinkage = 20 per cent., or 1-5th.

80 lbs, net cost \$50.00, as above.

```
80) 50.00 (62.50
    48 0
    ----
     200
     160
     ---
      400
      400
      ---
```

Some persons, however, figure in this way:

Example B.

100 lbs. at \$0.50 lb.
Shrinkage 20 per cent. = 1-5th.
\$0.50 + 1-5th (10 cents) = 60 cents.

Example A.—Correct method—net cost	62.50
Example B.—Incorrect method—net cost	60.00
Difference	2.50

This is a difference of 4 per cent., which, if it occurs in manufacturing a large amount of goods where rubber is the greater part of the compound, would make quite a difference in the profit.

SPECIFIC GRAVITY OF RUBBER.

THE following records of the specific gravities of different samples of India-rubber have been collected:

Best Para, taken in dilute alcohol (Ure)	0.941567
Best Assam, taken in dilute alcohol (Ure)	0.942972
Best Singspore, taken in dilute alcohol (Ure)	0.936650
Best Penang, taken in dilute alcohol (Ure)	0.919178
Caoutchouc (Julian)	0.920000
Crude caoutchouc of India (Adriani)	0.966800

Black caoutchouc (Adriani)	0.945200
Prepared from juice in pure state (Faraday)	0.925000
Determined by E. Soubeiran	0.935500
Determined by Payen	0.925000

H. L. Terry, F. I. C., gives the specific gravity of Para rubber and refers to Faraday's figures as being most correct.

Faraday's general analysis of the sap of the *Hevea* is:

Caoutchouc	30.70
Albuminous extractive and saline matter	12.93
Water	56.37

The specific gravity of the sap quoted was 1.012.

The crude rubber itself is made up of the following general composition: Carbon, 87.5; hydrogen, 12.5.

CHAPTER XIII.

PHYSICAL TESTS AND METHODS OF ANALYSIS OF VULCANIZED INDIA-RUBBER.

IT has long been the boast of expert rubber superintendents and manufacturers that they found little trouble in matching compounds. As a matter of fact, some of them are marvelously expert. Given a small sample of vulcanized rubber in a familiar line, with a knowledge of the price at which it must be produced, they are able in a majority of instances, by their knowledge of rubber and of compounding ingredients, to get a result that is apparently similar, and without much experimenting.

In certain instances, however, they fail, principally where a new product is brought in for matching, to which is attached an extraordinarily low price. The usual refuge in such a case formerly was the assertion that the manufacturer was losing money on that particular line of goods. But this has been so often disproved, and the sample found to be both an original and better compound, that this excuse is not often heard nowadays.

The factory expert gaged his sample, no matter how expert he might be, by purely physical rules. The smell told him what kind of rubber was used, whether Para or African, and usually whether reclaimed rubber was present. The strength and the weight of the sample gave him an indication as to the amount of adulteration. The color also had its suggestions as to material contained in it, but the knowledge thus shown often was very far from being exact.

Nor was the general result very much better when information was purchased from employes, or points secured through quizzing the supply men. The best course for the rubber superintendent to pursue, therefore, is to put his knowledge up against that of the expert chemist, when the two, working together, can usually match better than the original. It is better, if the chemist is familiar with the practical manipulation of rubber, for the unfamiliar chemist has in many cases brought science into considerable disrepute in the factory.

Certain rubber compounds, in spite of the most careful analysis by expert chemists, have remained, and probably will remain, profound secrets. For ordinary work, however, there ought to be no trouble in getting a fair analysis. The following descriptions of processes employed in the analysis of vulcanized rubber are given chiefly that the rubber superintendent who views chemistry as a dark and deep mystery may have some knowledge of what the chemist is about when he seeks his assistance. Before beginning on chemical analysis a few words more concerning physical tests may not be amiss.

In the case of many kinds of goods there is a great variety of appliances that form really valuable tests as to their durability, tensile strength, wearing quality, etc. As a rule, these aim to reproduce the work that the vulcanized article is obliged to endure in actual service. In rubber boots and shoes, for example, a machine is employed which bends the shoe exactly as it is bent when the wearer is walking, and at the same time gives a friction motion on the sole. This is run at a high rate of speed, so that a week's wear on a machine like this would correspond to a month of service in actual use.

A machine is also used for testing air-brake hose which counterfeits the swing and kinking motion that the hose gets in actual service. This is run at a very high rate of speed, and the hose which stands this sort of usage longest is supposed to be adapted to endure the longest time in actual use.

Tires, both pneumatic and solid, are tested by being put on a wheel rim and run what is equivalent to hundreds and thousands of miles over roughened surfaces upon which they are pressed by a lever carrying heavy weight. These mechanical contrivances are valuable in showing the severe usage that rubber will often stand, but none of them are exact parallels to absolute service, for as a rule they are more severe, particularly in the intense heating that may come to the rubber from high speeds and great friction.

Manufacturers and purchasers of rubber goods have also many simple and excellent tests for approximating the value of the rubber. In belt and hose covers and tubes, a bit of the rubber is cut from the fabric and stretched to show its tensile strength. The fabric is also pulled apart, and the integrity of the friction proved by the way it resists such separation. Rubber springs sometimes have been placed under a steam hammer which was allowed to drop upon them, the results being noted and that compound standing up longest being considered the best.

An English manufacturer following out this test, got some interesting, if not valuable, results. He took a piece of vulcanized India-rubber 1½ inches thick and with 2 inches area, and placed it under a steam hammer of five tons, which first rested upon the rubber without effect. The hammer was then raised two feet and dropped upon it without injury; then lifted four feet, when the cake was torn, but none of its elasticity was destroyed. More severe trials were then made. A block of vulcanized Inlia-rubber was placed between two cannon balls, with the whole power of the heaviest steam hammer employed; the iron spheres split the block, but the elasticity of the rubber still remained.

The ordnance department of the United States government some years ago inaugurated some very interesting tests of vulcanized rubber at the arsenal at Watertown, Mass., the results of which are appended:

NO. 1.

Applied Loads.	Mean Length.	Compression.	Compression Sets.	Middle Diameter.
Pounds.	*Inches.*	*Inches.*	*Inch.*	*Inches.*
0	5.72	....	....	6.10
1,000	5.32	.40	0.	6.38
2,000	4.84	.88	.10	6.72
3,000	4.47	1.25	.18	7.06
4,000	4.03	1.69	.29	7.48
5,000	3.70	2.02	.33	7.79
6,000	3.40	2.32	.37	8.12
7,000	3.14	2.58	.42	8.44
8,000	2.96	2.76	.39*	8.73
9,000	2.80	2.92	.51	8.92
10,000	2.68	3.04	.58	9.11
11,000	2.60	3.12	.52*	9.24
12,000	2.50	3.22	.60	9.42
13,000	2.45	3.27	.67	9.55
14,000	2.36	3.36	.73	9.68
15,000	2.31	3.41	.74	9.77
0	5.15	....	....	6.71

*Before these sets were taken the load on the rubber was reduced to 500 pounds, then increased to 1,000 pounds, and the sets then measured.

The second test was of new rubber gun-carriage springs, in which the compression sets were determined under the initial load, the end diameters approximate under load. The length of the rubber spring was 6.03 inches; the diameter 6.03 inches; the diameter of core 1.04 inches; the sectional area 27.71 square inches; and the weight 11 pounds:

NO. 2.

Applied Loads.	Length.	Compression.	Compression Sets.	Diameters. End.	Diameters. Middle.	Middle Diam. Under Initial Load.
Pounds.	*Inches.*	*Inches.*	*Inch.*	*Inches.*	*Inches.*	*Inches.*
500	5.87	0.	0.	6.03	6.12	6.12
1,000	5.70	.17	.02	6.03	6.24	6.15
1,500	5.51	.36	.03	6.03	6.35	6.15
2,000	5.34	.53	.06	6.03	6.48	6.18
2,500	5.13	.74	.07	6.03	6.64	6.18
3,000	5.00	.87	.07	6.10	6.76	6.18
3,500	4.81	1.06	.09	6.15	6.87	6.18
4,000	4.65	1.22	.08	6.16	7.00	6.18
4,500	4.50	1.37	.08	6.18	7.15	6.19
5,000	4.35	1.52	.02	6.29	7.26	6.19
5,500	4.20	1.67	.12	6.38	7.41	6.19
6,000	4.06	1.81	.19	6.43	7.55	6.19
6,500	3.95	1.92	.03	6.50	7.66	6.21
7,000	3.83	2.04	.15	6.64	7.77	6.21
7,500	3.70	2.17	.15	6.70	7.91	6.22
8,000	3.62	2.25	.15	6.78	8.02	6.22
8,500	3.52	2.35	.16	6.89	8.13	6.22
9,000	3.43	2.44	.16	6.96	8.24	6.23
9,500	3.35	2.52	.17	7.06	8.34	6.24
10,000	3.25	2.62	.17	7.25	8.46	6.25

The spring was then removed from the testing-machine, measured, and its length was 5.90 inches; middle diameter 6.08 inches. After it had rested 20 minutes the length was 6 inches, and the middle diameter 6.06 inches. It was then placed again in the testing machine and the figures on the following page taken.

When removed its measurements were: Length 5.86 inches; middle diameter 6.24 inches; end diameter 5.90 inches; the ends were concave, V. sine .06 and .08 inches. After six hours rest it recovered in length to 5.96 inches.

NO. 3.

Applied Loads.	Length.	Compression.	Compression Sets.	Diameter.		Middle Diam. Under Initial Load.
				Ends.	Middle.	
Pounds.	*Inches.*	*Inches.*	*Inch.*	*Inches.*	*Inches.*	*Inches.*
500	5.84	.03	.03	6.01	6.15	6.15
6,000	4.00	1.87	..	6.55	7.63	....
10,000	3.29	2.58	..	7.25	8.40	....
10,500	3.21	2.66	..	7.35	8.48	....
11,000	3.16	2.71	..	7.39	8.54	....
11,500	3.11	2.76	..	7.46	8.60	....
12,000	3.06	2.81	.17	7.55	8.67	6.26
13,000	2.94	2.93	..	7.74	8.86	....
14,000	2.86	3.01	..	7.86	8.97	....
15,000	2.80	3.07	.22	7.94	9.04	6.30
16,000	2.71	3.16	..	8.10	9.20	....
17,000	2.65	3.22	..	8.20	9.28	....
18,000	2.61	3.26	..	8.27	9.35	....
19,000	2.56	3.31	..	8.36	9.42	....
20,000	2.53	3.34	.30	8.43	9.47	6.37

In the next test the length of the spring was 6.06 inches; diameter, 5.97 inches; diameter of core 1.06 inches; sectional area 27.11 square inches; weight, 11 pounds.

NO. 4.

Applied Loads.	Length.	Compression.	Compression Sets.	Diameter.		Middle Diam. Under Initial Load.
				End.	Middle.	
Pounds.	*Inches.*	*Inches.*	*Inch.*	*Inches.*	*Inches.*	*Inches.*
500	5.90	0.	0.	5.97	6.07	6.07
1,000	5.75	.15	.02	5.97	6.16	6.07
1,500	5.59	.31	.02	5.97	6.27	6.08
2,000	5.41	.49	.06	5.98	6.38	6.10
2,500	5.25	.65	.05	6.02	6.48	6.10
3,000	5.05	.85	.09	6.05	6.62	6.12
3,500	4.90	1.00	.08	6.08	6.73	6.11
4,000	4.76	1.14	.10	6.14	6.88	6.12
4,500	4.61	1.29	.10	6.20	7.00	6.12
5,000	4.47	1.43	.11	6.25	7.11	6.12
5,500	4.33	1.57	.11	6.31	7.24	6.14
6,000	4.21	1.69	.12	6.37	7.32	6.15

The measurements when removed from the machine were: Length, 5.98 inches; middle diameter 6 inches; end diameter 5.97 inches. After it had rested 15 hours, it measured length 6.02 inches; middle diameter 6 inches; end diameter 5.96 inches. It was then placed again in the machine and tests were resumed.

NO. 5.

Applied Loads.	Length.	Compression.	Compression Sets.	Diameter.		Middle Diam. Under Initial Load.
				End.	Middle.	
Pounds.	*Inches.*	*Inches.*	*Inch.*	*Inches.*	*Inches.*	*Inches.*
500	5.90	0.	..	5.97	6.08	6.08
6,000	4.27	1.63	..	6.35	7.30	....
6,500	4.12	1.78	.08	6.38	7.41	6.11
7,000	4.00	1.90	.09	6.50	7.55	6.12
7,500	3.90	2.00	.10	6.57	7.62	6.14
8,000	3.82	2.08	.10	6.65	7.73	6.12
8,500	3.72	2.18	.11	6.75	7.82	6.14
9,000	3.62	2.28	.14	6.84	7.93	6.16
9,500	3.52	2.38	.14	6.93	8.03	6.17
10,000	3.45	2.45	.16	7.00	8.11	6.17

The spring was then removed from the testing machine and its measurements were: Length, 5.92 inches; middle diameter, 6.09 inches. Measurements after the spring had rested one hour showed: Length, 5.98 inches; middle diameter, 6.06 inches; end diameter, 5.95 inches. The spring was again placed in the machine and tests resumed.

NO. 6.

Applied Loads.	Length.	Compression.	Compression Sets.	Diameter.		Middle Diam. Under Initial Load.
				End.	Middle.	
Pounds.	*Inches.*	*Inches.*	*Inch.*	*Inches.*	*Inches.*	*Inches.*
500	5.88	.07	.07	5.97	6.13	6.13
6,000	4.09	1.81	..	6.41	7.47	....
10,000	3.46	2.44	..	7.00	8.12	....
10,500	3.38	2.52	..	7.09	8.22	....
11,000	3.34	2.56	..	7.15	8.27	....
11,500	3.27	2.63	..	7.21	8.36	....
12,000	3.17	2.73	.22	7.26	8.40	6.22
13,000	3.10	2.80	..	7.46	8.57	....
14,000	3.02	2.88	..	7.51	8.67	....
15,000	2.90	3.00	.26	7.66	8.70	6.27
16,000	2.84	3.06	..	7.84	8.95	....
17,000	2.79	3.11	..	7.90	9.02	....
18,000	2.75	3.15	..	7.97	9.06	...
19,000	2.70	3.20	..	8.05	9.13	....
20,000	2.68	3.22	.37	8.11	9.19	6.36

[THE PRECEDING TABLE CONTINUED.]

Applied Loads.	Length.	Compression.	Applied Loads.	Length.	Compression.
Pounds.	*Inches.*	*Inches.*	*Pounds.*	*Inches.*	*Inches.*
1,000	5.43	.47	11,000	3.05	2.85
2,000	5.10	.80	12,000	2.98	2.82
3,000	4.75	1.15	13,000	2.91	2.99
4,000	4.43	1.47	14,000	2.85	3.05
5,000	4.10	1.80	15,000	2.81	3.09
6,000	3.80	2.10	16,000	2.78	3.12
7,000	3.58	2.32	17,000	2.74	3.16
8,000	3.38	2.52	18,000	2.70	3.20
9,000	3.25	2.65	19,000	2.67	3.23
10,000	3.15	2.75	20,000	2.63	3.27

Time for loading three minutes. The spring was then removed from the testing machine and its measurements showed: Length, 5.81 inches; middle diameter, 6.25 inches; end diameter, 5.87 inches; ends concave, V. sine, .08 and .10 inch. It recovered in length to 5.93 inches after four hours' rest.

The French navy also inaugurated a series of tests for rubber belting which are of interest. The first test related to elasticity. Samples from the cover were first put into a steam vulcanizer for 48 hours, under a pressure of 5 atmospheres, which they should stand without losing their elasticity. The samples are then placed under a pressure of 85.5 pounds per square inch on the grating of a valve box, and given strokes at the rate of 100 per minute. They were expected to stand 9,100 strokes, while samples not tested by the steam should stand 17,100 strokes. Strips from the cover that had received the steam treatment, 6-10 of an inch square on cross section, and 8 inches long, fastened at each end and elongated 3.9 inches, were not expected to break when stretched to 8 inches more, this being repeated 22 times a minute for 24 hours. Strips that had not been treated to the steam bath, should resist the same treatment for 100 hours. These tests of course applied to high grade compounds only.

The analysis of vulcanized India-rubber should give the following information:

Amount of India-rubber,
Amount of India-rubber resins,
Amount of substitutes,

Amount of free, fatty resin, and mineral oils, resin, paraffine, and bituminous bodies,
Amount of sulphur of vulcanization,
Amount of sulphur and chlorine in substitute,
Amount of free sulphur,
Amount of mineral matters.

The mineral matters embrace metallic sulphides and oxides, inert mineral substances such as whiting and barytes, and substances imparting special properties, such as asbestos, graphite, pumice, etc.

According to Carl Otto Weber, Ph. D., F. C. S., and to Percy Carter Bell, F. I. C., F. C. S., Dr. Rob. Henriques has by his methods of analysis solved the problem that troubled the analysts more than any other, which was that of determining the amount of oil substitutes found in India-rubber compounds.

Dr. Henriques's methods are as follows: Fuming nitric acid to the amount of 20 c. c., is placed in a small dish covered with a funnel, through the stem of which 3 to 4 grams of rubber are slowly added. When the action has ceased, the dish is warmed gently on a water bath until the contents are of the consistency of a thin syrup. There is then added 4 grams of a mixture of 4 parts of sodium carbonate and 3 parts of potassium nitrate, after which it is carefully fused, and treated with dilute muriatic acid, then evaporated to dryness to render silica, if present, insoluble, redissolved by adding a little nitric acid, and, last, the sulphuric acid is precipitated with barium chloride. The residue of silica may contain sulphates of lead or of barium. Ammonium acetate dissolves the former.

In estimating the sulphur of vulcanization, and also the excess of sulphur, they must be separated from that present in the form of sulphates and sulphides. This is done in the following manner: The sample of rubber is dissolved in that fraction of ordinary petroleum which distills over at from 140° to 250° C., being kept in the solvent at a boiling temperature for two days. From 5 to 15 grams of the sample are placed in a weighed flask, and, after adding about 150 c. c., petroleum free from sulphur, all the inorganic matter is dissolved by heating the flask with reflux condenser at about 150° C. The subsequent processes are the filtering of the solution, the careful washing of the flask with hot petroleum, and the rinsing of both flask and filter with petroleum ether. Those substances insoluble in petroleum are determined by weighing on the tared filter at 110° C.

The sulphur in this residue which is easily determined, when deducted from the total sulphur of the sample, gives the amount of the free sulphur, and sulphur of vulcanization. If the rubber contains metallic oxides or carbonates, some of the sulphur may have been oxidized to sulphuric acid, and the results noted above may be too low.

The rubber substitutes in the compound are completely and easily soluble in alcoholic potash. The following is the manner of this analysis: From 3 to 5 grams of the rubber compound, finely divided, is boiled for about 8 hours in ten times its weight of alcoholic soda, 8 per cent. strong. The solution, diluted with water, is freed from the alcohol by means of a water bath, after which the residue on a weighed filter is washed, dried, and weighed. To determine the residue or ignition of the extracted residue, one gram is taken and the ignition performed in the presence of ammonium nitrate. If now the substance extracted from the rubber is free from chlorine, it may either consist of free oil, or be derived from black rubber substitute. In the latter case, it must contain at least 10 per cent. of sulphur, but in the former, only traces of sulphur will be present. An estimation therefore of the chlorine and of the sulphur in the alcoholic extract determines the presence of white substitute, black substitute, or sulphur.

In using caustic alkali a certain amount of the alkali will be retained, the amount of which must be determined, if correct figures are to be secured. Repeated washings in dilute muriatic acid remove this, and allow of its determination.

The following data are necessary in the analysis of vulcanized rubber containing substitute or oil: (1) The total sulphur; (2) the total ash; (3) the weight of the substance after extraction with alcoholic soda; (4) the sulphur, the ash, and the sulphur in the extracted fatty acids all to be found in the third substance. Also, the weight of the substance after extraction with alcoholic soda. From 1.5 to 2 grams of substance are used, the extraction being twice repeated, each boiling being from two to three hours. The quantity of rubber dissolved by the alcoholic soda is deducted from the weight of the total extract. This correction averages 2.5 per cent.

From the above figures, the percentage of rubber and fatty acids may be calculated by equations, which read:

$$\text{Rubber} = \frac{100}{97.5}\ (\text{Weight of substance after extraction of alcoholic soda} - \text{its sulphur} - \text{its ash}).$$

The fatty acids from this equation:

Fatty acids = 100 — (total sulphur + total ash + percentage of rubber found from the foregoing equation).

The sulphur contained in the rubber substitute is represented by assuming that quantity to be about equal to that of the fatty acids in white substitute and about 1.5 per cent. larger than the quantity of fatty acids in brown substitute. The difference between the total sulphur and the sulphur in the substitute is the sulphur of vulcanization. Asphalt being often present in rubber compounds, by first dissolving the free sulphur by treatment with alcoholic soda, and then dissolving the asphalt out by means of nitrobenzene, it is easily determined. The presence of mineral oils, paraffine, and resins are the only things that interfere with this means of extraction.

The following tests are credited to C. A. Lobuy de Bruyn:

1. Extract Test.—(Henriques's method).—Three grams of the finely divided sample when boiled for six hours with 50 c. c. of a 6 per cent. alcoholic solution of caustic soda should not lose more than 8 per cent., the loss to be calculated upon the organic substance of the sample. The extract should contain sulphur and rubber resins.

2. Dry Heat Test.—Two grams of the finely divided sample are heated to 135° C. for two hours. When cold the sample should not have suffered any alteration and should show a loss of weight not exceeding 1.5 per cent.

3. Moist Heat Test.—A small piece of the sample is sealed in a glass tube half filled with water. The tube is then heated to 170° C. for four hours. The sample should not be affected by this treatment.

4. Ash.—About 1 gram of the sample is fused, decomposed,

and partly ignited over a small flame in a porcelain crucible. The heat is then increased and ignition completed.

Dr. C. Reinhardt, in *Dingler's Polytechnisches Journal,* writes as follows on the analysis of vulcanized India-rubber: "The determination of the ashes is effected by gradually heating in a covered crucible .0182 ounce of the product until the cessation of gaseous liberation. The calcination is finished in an open crucible, care being taken not to heat too much, so as to avoid the losses due to the volatilization of the substances composing the ashes. To determine the proportion of mineral substances (with the exception of sulphur) .0182 ounce of India-rubber fragments is moistened with 1.2 cubic inch of D nitric acid (= 14. and heating takes place in a water bath for five to seven minutes, until complete dissolution ensues. Dry evaporation takes place in the water bath, followed by moistening with hydrochloric acid and dissolution in water. The residue is formed of sulphate of barium and silica acid; the quantitative analysis of the substances contained in the liquid (oxide of zinc, lime, magnesia, oxide of iron, and alumina) being made according to the usual methods. To determine the total of sulphur there is treated .0357 ounce of the product (while heated) with 1.2 cubic inches of nitric acid; chlorate of potash being gradually added until oxidation is complete. After evaporation and dissolution in water, with the addition of hydrochloric acid, follows precipitation. Then takes place, the quantitative analysis of the sulphuric acid by the chloruret of barium and of the remainder of the sulphuric acid in the insoluble residue of sulphate of baryta. It is possible to determine the quantity of sulphur added for the vulcanization by burning the product in a current of oxygen at a low temperature by passing the vapors across hydrochloric acid containing bromine, and by analyzing quantitatively the sulphuric acid formed in the condition of sulphate of baryta. The India-rubber can likewise be distilled in glass tubes and the quantity of sulphur in the distilled liquor can be ascertained."

Dr. Weber's exceedingly valuable article printed in the *Journal of the Society of Chemical Industry* is probably the most comprehensive treatment that the subject of the analysis of vulcanized rubber has yet received. The steps in that analysis are thus shown:

SUMMARY OF WEBER'S METHODS OF ANALYSIS.

<table>
<tr><td colspan="5">I. Acetone (10 runs in Soxhlet tube).</td></tr>
<tr><td rowspan="5">Fatty and Mineral Oils, Resins and Free Sulphur.</td><td colspan="4">II. Boiling Alcoholic Soda (8 per cent).</td></tr>
<tr><td rowspan="4">Rubber Substitutes.</td><td colspan="3">III. Cold Nitrobenzole.</td></tr>
<tr><td rowspan="3">Asphaltum.</td><td colspan="2">IV. Boiling Nitrobenzole (Soxhlet tube).</td></tr>
<tr><td rowspan="2">Rubber and Sulphur of Vulcanization.</td><td>V. Residue.</td></tr>
<tr><td>Mineral matters and free carbon.</td></tr>
</table>

The rubber substitutes are determined by extracting in alcoholic soda solution and asphaltum by cold nitrobenzene, both of these methods being Henriques's. The rubber is separated by extraction with boiling nitrobenzene in the Soxhlet tube. Starch is dissolved out by boiling water. The mineral and carbonaceous matters are determined in the final residue. The matters in the acetone extract, the rubber and mineral matters are determined by weighing after evaporation. Substitutes and asphaltum are best determined in the loss of weight operated upon.

Of the various forms of sulphur occurring in rubber, the determination of free sulphur and sulphur of vulcanization, is of great importance. The estimation of the free sulphur is made in the acetone extract. Not all the sulphur in this extract is free, as the presence of rubber substitutes in the sample means that the extracts will contain sulphides of the fatty acids, also the sulphides produced by the action of free sulphur on the resins always ~~found in rubber~~ To estimate the sulphur in the acetone extract, add 20 c. c. of a solution of pure sodium sulphide and caustic soda and heat the mixture on a water bath for an hour. Dilute the solution with warm water, and precipitate the fatty acids by adding a slight excess of barium hydrate. Filter, wash, and make up the filtrate to 300 c. c. and estimate the sulphur in an aliquot part.

In determining the sulphur of vulcanization, the free sulphur must first be removed, and for this purpose, the acetone extract answers very well. In every case the sulphur of vulcanization should be estimated direct. The solution of rubber in nitrobenzene is therefore distilled under reduced pressure. The flask containing the non-volatile residue is then dried at 140° C., and then oxidized with fuming nitric acid. When the residue has finally dissolved, the solution is poured into a platinum dish, the flask being rinsed with warm nitric acid. The residue is then evaporated on the water bath, fused with carbonate of soda, dissolved in water, oxidized with bromine, acidulated with muriatic acid, and the sulphur precipitated with barium chloride. The sulphur in the asphaltum which is in the cold nitrobenzol solution is determined in a similar manner.

The *India-Rubber and Gutta-Percha Trades Journal* thus briefly summarizes processes for analyzing vulcanized rubber: "The analysis of crude rubber does not offer great difficulties. The sample has carefully to be taken, which is best done with the help of rollers, as used in rubber works. While kneading the rubber on the rollers, the rubber is mechanically purified by a water spray, and the loss in weight ascertained. Of the dried substance, 5 or 10 grams are extracted by a Soxhlet apparatus with acetone for several hours, when the rubber resins pass into solution; both the residue and ashes are then determined. Finished articles can generally be filed into friable powder This is digested with alcoholic soda lye, filtered, and washed with hot alcohol; the residue is boiled with water, the liquid always being passed through the same filter, then with hydrochloric acid, one filter being used, quickly dried, and weighed. The residue would still contain the bound sulphur, silicates, sulphates of barium, etc. What remains when sulphur and ashes have been allowed for, may be put down as rubber."

CHAPTER XIV.

GUTTA-PERCHA—ITS SOURCES, PROPERTIES, MANIPULATION, AND PRINCIPAL USES.

GUTTA-PERCHA, which was introduced into Europe from Singapore in 1843, was for awhile confounded with India-rubber, from which it differs in some very important particulars. It becomes soft and plastic on immersion in hot water, retaining the shape then given it on cooling, whereupon it becomes hard, but not brittle like other gums. India-rubber, on the other hand, does not soften in hot water, and retains its original elasticity and strength almost unimpaired. The water, as such, exercises no softening action on Gutta-percha, the effect being purely one of temperature, which may equally well be produced by hot air, only somewhat more slowly. The degree of heat required depends upon the quality of the material, but even the hardest kinds become plastic above 150° F. Heated in air considerably above the boiling point of water, Gutta-percha decomposes and finally ignites, burning with a luminous smoky flame and emitting a pungent odor resembling that from burning rubber. If heated in a vacuum, gaseous and liquid products are obtained similar to those resulting from the distillation of rubber. The liquid which distils over consists chiefly of hydrocarbons of the terpene series, which form an excellent solvent for caoutchouc. The two most important are *isoprene* and *caoutchine,* which are identical with the liquids by the same names obtained from India-rubber. Since these products can also be obtained from other sources, Dr. Eugene Obach and others have observed that they may yet form a stepping-stone in the synthetical production of India-rubber and Gutta-percha from the lower terpenes.

A curious physical characteristic of Gutta-percha is that when it has been softened in water, although it is so plastic that it will reproduce the most delicate impressions, it will bear blows from hammers or allow itself to be thrown against a stone wall without being at all marred. The reason for this is that it contains a large amount of air. By placing the Gutta-percha under a bell jar immersed in mineral oil, when a vacuum is produced, a large amount of air is evolved from the gum, and it will be found to have lost the property of hardening on cooling, its substance being like a tough greasy leather.

Nowhere on the globe have genuine Gutta-percha trees been found outside of a rectangular area embracing portions of the Malay peninsula, Borneo, Sumatra, and some adjacent smaller islands. Strange to say, the occurrence of these trees has not been established—though they may yet be discovered—in Java, the Celebes, or the Philippines. These trees belong to the natural order *Sapotaceae;* the principal genera and species will be noted further on.

According to Payen's analysis, verified by later chemists, Gutta-percha contains three components: (1) a substance insoluble in cold and in boiling alcohol, which he termed pure *gutta;* (2) a crystaline white resin, soluble in hot, but not in cold alcohol, which he called *albane;* (3) an amorphous yellow resin, which he named *fluavile.* Pure gutta is insoluble in ether and light petroleum spirit at ordinary temperatures, whereas both albane and fluavile dissolve readily in them. Gutta possesses all the valuable qualities of Gutta-percha, but in a much enhanced degree; it becomes soft and plastic on heating, and hard and tenacious on cooling without being in the least brittle. But the resins themselves are either soft at ordinary temperatures, or, when hard, quite friable. It is, therefore, gutta which forms the useful constituent of Gutta-percha, and the resins are only accessory components, which, although admissible, and perhaps even desirable in a comparatively small amount, yet have a decidedly detrimental effect when they preponderate. Hence, in order to determine the technical value of a sample of Gutta-percha, it is necessary first to learn the relative proportion or ratio between gutta and resins. There must also be taken into account the *water* enclosed in the mass, and the coarser impurities—wood fibers, bark, sand, etc.—which are described as *dirt.* These components represent the loss or waste to the manufacturer.

While the relative proportion of gutta and resins forms an important criterion for estimating the commercial value of a sample, it is not in itself sufficient. Although the analysis of two different specimens may give the same result, the physical and mechanical properties, and, most important of all, the durability, may differ widely, owing to a difference in their molecular constitution. It will thus be seen that there are guttas and guttas. In addition to the qualitative analysis, it is necessary to scrutinize the gutta itself, which requires much judgment and experience. Analyses have been made of specimens which contained eight times as much gutta as resin; others contained about an equal amount of both, and in others still the amount of resin was three times that of gutta. Samples in which the percentage of resin reaches that of gutta, or surpasses it, are of a decidedly inferior description. These differences are due doubtless to the fact that the Gutta-percha of commerce is derived from trees of various species, and also in part to the treatment which the gum receives at the hands of the gatherers, who are suspected of mixing the product of different trees, to say nothing of adulterations of a more debasing character.

The commercial classification of Gutta-percha is less satisfactory than that of India-rubber, since no standards have become fixed in the markets. While Para rubber, for instance, may be bought and sold by means of established designations, "Islands fine," "Upriver fine," and the like, no such practice exists with regard to Gutta-percha. Since all transactions in the latter are based upon samples, trade names and brands are little considered. However, "Macassar" and "Banjermassin," which are the names of districts producing Gutta-percha, were used formerly to indicate the highest quality, while "Sumatra" sorts were supposed to be less valuable, and Borneo the lowest of all. In a sense these designations have become merely commercial, no longer affording any indication of the origin of the Gutta-percha. At the same time, "Macassars" and "Banjermassins" might vary with every new arrival, so that one was not certain, in buying one of the sorts named, to obtain particularly good Gutta-percha; it might have been the very opposite.

Innumerable sorts appear in the Singapore market—which is the center of the Gutta-percha trade—but Dr. Obach selected twelve of the principal brands as typical of all the rest, and divided them into four groups, for convenience in comparison, the best being named first. They are as follows, the designations being derived either from the countries of their origin or from the places of export:

I.	1. Pahang—from the Malay peninsula. 2. Bulongan red—from Macassar, Borneo. 3. Banjer red—from Banjermassin, South Borneo.
II.	4. Bagan goolie soondie—from Borneo. 5. Goolie red soondie—from Serapong, Borneo. 6. Serapong goolie soondie—from Serapong, Borneo.
III.	7. Bulongan white—from Macassar, Borneo. 8. Mixed white—from Borneo. 9. Banjer white—from Banjermassin, South Borneo.
IV.	10. Sarawak mixed—from Borneo. 11. Padang reboiled—from Sumatra. 12. Banca reboiled—from Banca.

Group I comprises the three best kinds, derived from trees of the genus *Dichopsis* (known in continental Europe as *Palaquium*). Group II comprises three kinds of the second order, derived probably from the genus *Payena*. Group III embraces the so-called "white gutta," of second and third grade, mostly of uncertain origin, but probably from *Dichopsis polyantha*. Group IV is made up of mixed materials, two of them being what is termed "reboiled" (an operation performed by the Chinese traders, who buy up odd lots, soften the materials in hot water, and make them into a more or less homogeneous average mixture). The "Sarawak mixed" lots mostly represent a very useful second-class material; the "reboiled" is decidedly inferior. This classification is based upon the results of 751 analyses of mixed lots, representing over 5,000,000 pounds of raw Gutta-percha, made by Dr. Obach, with a view to arriving at the relative proportions of gutta, resin, dirt, and water contained. The cleanest kind is the "Serapong soondie," which contains only 3½ per cent. of dirt, but it is rather wet, having more than 25 per cent. of water. One of the least favorable materials is "Banjer white," which contains 33 1-3 per cent. of water and 15 per cent. of dirt, making in all nearly 50 per cent. of waste. When a raw material is very dirty and wet, it is noticeable on cutting the blocks open, and this is now the rule in the Singapore market. The blocks are then sorted out into several grades (two or three, sometimes more) according to their appearance, and valued accordingly.

A grade of Gutta-percha which is nearly white in color and very brittle is apt to contain a large percentage of resin, which, as already explained, renders it of little value. In explanation of some of the terms in the preceeding classification, it may be said that Gutta-percha is obtained principally by cutting down the trees and ringing the bark at intervals of 12 to 18 inches along the trunk. The milky sap soon fills the grooves cut into the bark, and, in the better varieties, soon coagulates, when it is scraped off with a knife. In the case of inferior sorts, the milk requires more time to curdle, and has to be caught in receptacles placed under the tree. The collected milk is then gently boiled, either by itself or with the addition of water. The material obtained without the use of water is called a *goolie,* the other a *gutta;* but the two kinds are often mixed together. The goolie is more compact than the gutta, and has a dough-like smell. The word soondie is derived from the Malay term "Gutta-sundek," which is applied to the product of trees of the *Payena* species already referred to.

The processes employed by manufacturers for cleaning raw Gutta-percha are either *mechanical* or *chemical.* Those of the first class will first be considered. Generally speaking, the raw Gutta-percha is either first cut up in a slicing machine and then softened in hot water, or the lumps are placed directly in hot water and the soft material transferred to the washing machine. There it is washed with hot water for a longer or shorter time, and then passed through a strainer. Next, as a rule, it is washed once more, then put into a kneading or masticating machine, to consolidate it and remove the mechanically enclosed water, and finally it goes to the rolling mill, to be made into sheets.

The slicing machine or chopper now used is pretty much the same as that proposed by Charles Hancock, of England, in his patent (No. 11, 575, O. L.) of 1847, except that it is is provided with a greater number of fluted and serrated knives, instead of only three plain ones, fixed in the slots of a heavy iron disc. The blocks of Gutta-percha are packed into a trough and then forced against the rotating disc, the knives in which cut the material into thin slices.

The washing machine consists of an iron roller of star-shaped section, enclosed in a cylindrical shell provided with one or two projections, or ribs, against which the Gutta-percha is forced in going round. The cylindrical shell is enclosed in a large iron case, filled with water, which is heated by means of direct steam. The dirt, as it is washed off, falls through the lower part of the cylindrical shell into the outer case, whence it is drawn off once in a while. This machine is developed from that described in the English patent of R. A. Brooman (No. 10,550, O. L.)

The Gutta-percha leaves the washing machine in a plastic state and passes to the straining machine—a strong iron cylinder with a perforated bottom, on which a number of discs of fine wire gauze have been placed. It has a piston which is driven home by hydraulic power, at a pressure of 1,500 to 2,000 pounds per square inch, squeezing the soft Gutta through the meshes of the gauze.

The kneading machine or masticator resembles the washer, except that the roller is smaller in diameter, and the flutings are more numerous and not so deep. The Gutta-percha is kept hot during mastication and the water escapes in the form of steam through openings at the top.

The mixing machine, introduced by Paul Pfeiderer, is similar to that used in the India-rubber, linoleum, and other similar industries. It is provided with peculiarly-shaped blades, working against one another. The machine is used for mixing the various sorts of Gutta-percha, in order to obtain a material of any requisite properties, and also for blending Gutta-percha with pigments or other ingredients. The rolls can be heated by steam, but heat is developed by the kneading process itself, and care must be taken not to overheat the material.

The Gutta-percha is next rolled into sheets, usually between $\frac{1}{8}$ and $\frac{1}{4}$ inch, and cut into lengths of 5 or 6 feet, and stacked away for use. The rolling machine takes the material from the mixer and squeezes it between parallel rollers, running it back and forth until it is cool and hard enough for cutting up.

The average percentages of waste, shown by numerous analyses of the twelve brands of Gutta-percha catalogued on a preceding page, are about as follows:

Pahang	34	Bulongan white	43
Bulongan red	35	White mixed	35
Banjer red	44	Banjer white	47
Bagan goolie soondie	32	Sarawak mixed	44
Goolie red soondie	27	Padang reboiled	44
Serapong soondie	36	Banca reboiled	29

The difference in the quality of various brands of Gutta-percha, measured by the relative proportions of gutta and resin, has already been mentioned. Of the sorts mentioned above, "Banca reboiled" shows a comparatively small loss in cleaning, but it is the least valuable on the list, being low in gutta, whereas "Pahang," though losing more in the cleaning process, is by far the most valuable sort in the market, because so rich in gutta. Gutta-percha imported in recent years loses more in cleaning than formerly; Dr. Obach, in 1898, estimated the loss as almost twice as great as formerly.

The chemical washing process was suggested by Charles Hancock, in an English patent, in 1846. He steeped raw Gutta-percha, cut into small pieces, in a solution of caustic alkali or chloride of lime, to neutralize the acidity and remove any unpleasant odor. His experiments showed that the alkaline treatment not only reduced the percentage of dirt—that is, it was better cleaned than by the mechanical process—but lessened the capacity of the Gutta-percha for retaining mechanically enclosed water. But the treatment with chemicals requires great care and judgment, and thorough subsequent washing with water; otherwise the material will be rendered perishable.

Chemicals were also used by Obach for hardening Gutta-percha. The really valuable constituent of Gutta-percha being the gutta, the more a sample contains of the latter, the better it is, provided the gutta itself is of a good description. For certain purposes it is advantageous to improve the hardness and other mechanical properties of Gutta-percha, and this can be done by extracting the resin with a suitable solvent, which leaves the gutta itself intact. The raw Gutta-percha is first chopped and thrown on drying platforms gently heated from below by steam pipes. Or the pieces may be thrown into a rotating drum heated by currents of warm air. They then go to a series of tanks in which petroleum spirit is used as a solvent for the resin. The spirit becomes charged with the resinous matters, and the resulting solution is distilled off, after which the material remaining is masticated as in the case of any other Gutta-percha. A specimen treated by this process will remain quite hard under a temperature which will render other specimens soft and plastic.

Other liquids may also be used, as ether, and a saturated solution of carbon disulphide in alcohol.

Instead of removing impurities from Gutta-percha by washing it either with water or an alkali, this can be done by dissolving the material into a suitable liquid, straining or filtering the solution, and then evaporating the solvent. Carbon disulphide has been used as the solvent, but with the effect of rendering the Gutta-percha perishable.

Recently an article known as Green Gutta-percha has been offered to the trade, being extracted from the leaves of the trees. Several systems for extracting Gutta-percha from leaves have been described. That of Dieudonne Rigole involves the use of carbon disulphide; that of Eugene Serullas the use of hot toluene as a solvent, after which the Gutta-percha is precipitated by means of acetone, instead of distilling off the solvent; and that of Obach the use of light petroleum spirit as a solvent for leaves that have been previously crushed between rollers, the gum being reprecipitated from the solution on cooling below 60° F. The author of each process has devised apparatus for its operation.

Many trees produce gums which have been experimented with in the hope that they would prove good substitutes for Gutta-percha, but none has proved of value except the "bullet" tree, which yields Balata. The gutta contained in Balata is very strong and tough, being of excellent quality; but the percentage of resin is large, and the material can be regarded as a substitute only for second-class, or perhaps even third-class, Gutta-percha. Balata is somewhat more flexible than Gutta-percha containing an equal amount of resin, which appears to be due to the softness of the resinous constituents. On becoming heated Balata behaves much like ordinary Gutta-percha. If plunged into boiling water it becomes quite soft and plastic. If next immersed in cold water, it slowly hardens again, but still remains flexible and elastic, showing no signs of brittleness. Analyses of specimens of Balata from British Guiana, obtained from the London docks in 1889-94, showed an average loss of 13.8 per cent. of water, and 9.9 per cent. of dirt, or a total of 23.7 per cent. of waste. The respective percentages of gutta and resin were 41.4 and 34.8.

The specific gravity of cleaned Gutta-percha is practically

the same as that of water, though varying with the relative proportion of gutta and resin, becoming lower as the percentage of resin increases. It may be affected, also, by the constitution of the resin and also of the gutta. The softening temperature of Gutta-percha depends entirely upon the ratio of gutta and resin. A specimen of which 60 per cent. was resin was softened at the temperature of 48° C. to the same extent as another specimen, containing only 2½ per cent. of resin, for which a temperature of 55° C. was required. The time for the material to become hard again, after having previously been softened in hot water, depends in a like degree upon the proportion of gutta and resin. But the principal mechanical property of Gutta-percha with which the manufacturer has to deal is the tensile strength. A specimen having 45 per cent. of gutta and 55 per cent. of resin will break under pressure of 770 pounds to the square inch, whereas for another specimen, after most of the resin has been extracted with petroleum spirit, nearly twice that breaking strain would be required. As for the elongation of Gutta-percha—*i. e.,* the extent to which it will stretch before breaking—it is also affected by the percentage of resin, being in the last two cases, for instance, 490 and 500 per cent., respectively, but it also depends on the nature of the gutta.

The earliest practical use of Gutta-percha was for surgical appliances—for bandages, splints, and receptacles for vaccine virus. It is used for ear trumpets; for the handles of surgical instruments, as it affords a firm grip and is preferable to wood for antiseptic reasons; in medicine, in the form (1) of a very thin tissue, (2) of sticks, and (3) of a 10 per cent. solution in chloroform; for chemical purposes, in the form of tubes, pumps, syringes, bottles, and the like, and for ladles and tubes for handling caustic alkalies and corrosive acids and liquids in chemical works; and for mechanical purposes, as rings and cups for pumps and hydraulic presses and for driving-bands (belting). For the later purpose Balata is also used largely, interposed between canvas; such belts can be joined by means of a solution of Balata or Gutta-percha in carbon disulphide. Another application of Gutta-percha is that for taking impressions of medals, and also of the interior of large guns. Gutta-percha is also modelled into ornaments in the shape of the leaves and petals of flowers, this being done by working the gum by hand in hot water with one or two simple iron tools. Such ornaments are often applied to the decoration of jars made of semi-porous ware, the whole being painted afterward.

But the most important application of Gutta-percha is in the insulation of submarine and subterranean cables. Dr. Werner von Siemens first proposed Gutta-percha for insulating purposes in 1846, and in the next year he designed a screw-press, for the seamless covering of wires with that material, which is still in existence, while the principle of the press is still adhered to. Gutta-percha has been found to be very permeable to the X-rays, and it has been proposed to utilize this property to examine Gutta-percha-covered wires for the detection of defects in the copper conductor, particularly in "joints," or for finding air-bubbles. The X-rays may also be used for the detection of large foreign bodies in the raw Gutta-percha. Up to the end of 1896 no less than 184,000 miles of commercial submarine cables had been laid, embodying the use of Gutta-percha of a weight estimated at 16,000 tons. Another 100,000 miles of cable had been laid by the various governments for military defense, which would require 8,000 tons more, or a total of 24,000 tons for submarine cables. A further allowance must be made, for underground cables, street wires, etc., of 8,000 tons. The length of Gutta-percha-covered wires under the streets of London alone is 17,000 miles, corresponding to 375 tons of Gutta-percha.

The electric properties of Gutta-percha depend chiefly on the nature of the gutta and to a less extent upon the resin; but only very slightly on the relative proportion of these two components. They depend also upon the nature and amount of the impurities and on the water. The insulation resistance and inductive capacity are little affected by the extraction of the resin. The insulation should be as high as possible, and the inductive capacity, for most purposes, as low as possible, but whereas the latter is mostly associated with other good qualities of the material, such is not always the case with a high insulation. A third electric property is called dielectric strength, or resistance to piercing by high voltages. A thickness of a little over $\frac{1}{8}$ inch of Gutta-percha breaks down with 40,000 volts, and one of about 1-10th inch with 28,000 volts.

Gutta-percha hardened by the extraction of its resin is used chiefly in the manufacture of golf balls. Gutta-percha for this purpose should be tough, elastic, and not brittle at low temperatures; it should be specifically lighter than water, in order not to sink if dropped accidentally into a ditch. It is requisite that the proper grade of raw material be chosen and that the resin be extracted as completely as possible. To test the elasticity of golf balls, a machine is used, consisting (1) of a perpendicular scale, divided into feet and tenths; (2) a clip, at the top, for holding the ball to be tested; and (3) an iron plate at the bottom. The object is to measure the rebound of the ball, when released from the clip and falling upon the plate. A ball made of Gutta-percha, of which 25 per cent. was resin, rebounded only to the point on the scale marked 30; a ball containing only 10 per cent. of resin rebounded to 45; and still another, having only a small percentage, rebounded to 60—the highest point reached. A ball of Balata, having the resin thoroughly removed, rebounded to 59.

Some figures will give an idea how greatly the physical and mechanical properties of Gutta-percha are affected by the extraction of the resin. Carefully selected specimens of a medium quality were cut fine and intimately mixed, and then divided into two portions. One portion was next washed in the ordinary way with water; the other treated with petroleum spirit until nearly all the resin had been extracted. The two specimens showed the following analyses:

	Gutta.	Resin.	Dirt.	Water.	Total.
Cleaned in ordinary way	54.7	39.4	2.7	3.2	100
Same material, hardened	93.0	2.8	2.5	1.7	100

The different physical and mechanical properties of the two specimens are indicated in the next comparison:

	Ordinary.	Hardened.
Temperature when commencing to soften	37.7°C.	57.2°C.
Temperature when commencing to harden	58.8°C.	91.1°C.
Time of hardening	17 min.	45 sec.
Tensile strength—pounds per square inch	1592	5662
Elongation—per cent	360	285

The electrical properties, on the other hand, are but little affected, the insulation being practically the same as before, and the decrease of specific inductive capacity is probably due to the smaller percentage of water in the hardened material.

The principal cause of the destruction of Gutta-percha is the absorption of atmospheric oxygen, which alters the gutta and produces a brittle resin of quite a different nature to that originally present in the material. This destructive oxidization is greatly assisted by light, and by other causes—for instance, by any action tending to make the material porous, such as alternate wetness and dryness, the presence of substances which exercise a solvent action on Gutta-percha as a whole, or any of its components. Certain alkaline substances and decaying organic matters also appear to act injuriously, but frequently it is impossible to assign a definite cause for the decay of Gutta-percha. It is, however, not merely manufactured Gutta-percha which undergoes these destructive changes, for raw material of the very best kind succumbs in time to the combined action of light and air. On the other hand, specimens of Gutta-percha are in existence which, after proper means of protection, have remained in good condition for more than fifty years. Complete immersion in water affords a good protection, for which reason submarine cores of Gutta-percha are more safely placed than underground wires. Another way of excluding the air, to some extent, is to varnish the Gutta-percha articles. When Gutta-percha is oxidized it becomes porous and full of cracks. If it is used for insulating wires, the insulation fails at such places, since the moisture penetrates the pores and fissures and establishes an electric contact with the conducting wire.

Some compounds containing Gutta-percha are very useful for different purposes, and a specially useful one, consisting of a mixture of Gutta-percha, colophony, and Stockholm tar, is known as "Chatterton's compound." It is used largely in connection with the manufacture of Gutta-percha-covered wires, as a binding material between the copper conductor and the Gutta-percha covering, or between the different layers of Gutta-percha on the core.

Willoughby Smith patented the following compound for insulating wires: One-fifth by weight of Stockholm tar and about the same weight of resin are put into a vessel with a jacket (or, preferably, a series of pipes) heated by steam; when properly melted the whole is passed through a wire gauze strainer "into another vessel similarly heated"; three-fifths by weight of Gutta-percha, having by preference, been previously cleansed in the ordinary way, and reduced into thin pieces or shreds, is then put into the heated vessel and mixed with the resin and tar. In this second vessel are stirrers, which are used to mix the whole uniformly.

Leonard Wray's cable compound was made of 1 part Gutta-percha, 4 parts India-rubber, 2 parts shellac, 2 parts flower of glass. This was used for underground wires.

Gaullie combined Gutta-percha with Roman cement by means of animal gall, forming a plastic material, capable of being stamped and molded.

Cooley mixed Gutta-percha with resin oil under heat, then mixed in carbonate of soda with roasted starch. To this compound he added asphalt to make it harder, or hyposulphite of lead, to make it softer. He also made a great many Gutta-percha compounds in which salts were present. These he steeped in water after mixing until they became soft and flexible.

Charles Macintosh made a compound for telegraph wire from Gutta-percha, naphthaline, and lampblack.

Charles Hancock boiled Gutta-percha in muriate of lime, passed it between heated cylinders, sifting the surface with rosin, in the production of a compound for complete insulation. Another of his compounds was made of Gutta-percha, shellac, and borax. He also made Gutta-percha sponge by mixing with it carbonate of ammonia or alum and applying heat. He also made a hard Gutta-percha which was similar to vulcanite by mixing it with sulphur, putting it in molds and keeping the compound at a high temperature for several days.

Duncan invented a great many compounds for Gutta-percha cement, many of which are now in general use. One suggestion of his was the mixing of Gutta-percha with Canada balsam and shellac, the resultant compound being a good cement capable of standing considerable heat and in no danger of becoming greasy on its surface.

Robert Hutchinson claimed that he was able to render Gutta-percha less liable to oxidize, to improve its elasticity, increase its tenacity, and diminish its liability to become sticky or tacky, by compounding it with lanichol or wood cholesterin. (See Lanoline). Forster deodorized Gutta-percha by mixing with it essential oil, orris root, or gum benzoin.

Liquid Gutta-percha is Gutta-percha dissolved in chloroform, to which a little carbonate of lead is added in the shape of a fine powder. After agitation, the mixture is set aside until the insoluble matter has settled. The clear liquid is then decanted.

Spill, in order to prevent Gutta-percha that had been vulcanized from being attacked by grease, treated it to a solution of melted beeswax, hardening this coating with an infusion of nut galls. Godefroy mixed Gutta-percha with powdered cocoanut shell, claiming that it would stand a higher degree of heat, and was considerably more elastic. Day mixed pipe clay with Gutta-percha that is being vulcanized in order to prevent its sponging.

The vulcanization of Gutta-percha, in spite of a common impression to the contrary, is something that can be easily accomplished, and is analgous to the vulcanization of India-rubber. It can be done by mixing with free sulphur or sulphides that contain free sulphur, or by the use of chloride of sulphur. As the Parkes mixture attacks Gutta-percha very easily, the dipping for vulcanization must be very quick, the article being then allowed to remain in the air for some hours. The second dip can be a little longer, as the surface is less easily attacked than before. The vulcanized product is quite hard and will stand a high degree of heat. Chloride of sulphur mixed with bisulphide of carbon can also be incorporated in a solution of Gutta-percha and bisulphide of carbon, with the result that the Gutta-percha will be thoroughly vulcanized.

The late Robert Dick, of Glasgow, who was a successful manufacturer of Gutta-percha articles in the mechanical line, produced many vulcanizable compounds of Gutta-percha of great value, some of which follow. He claimed that his compounded Gutta-percha retained the good qualities of the gum; that is, that it was homogeneous and plastic at a moderate heat, but tough and hard at ordinary temperatures, and that it was just as valuable afterwards for mixing and molding over again.

Compound No. 1 is described as the hardest and toughest, and may be used, in place of leather and vulcanized India-rubber, for tires, belts, pulley coverings, horse shoes, etc. No. 2 is softer and more elastic, and suitable for soles and heels of shoes, wringer rolls, springs, playing balls, mats, etc. These goods are mixed in the usual way, and vulcanize in the masticator, but not enough to take away the plastic qualities of the Gutta-percha. For treating this compound, a special masticator was devised by Mr. Dick, the rolling cylinders being hollow, and a Bunsen gas burner inserted through one end of the hollow axle, while the gases pass off at the other, thus heating both roller and mixture. The outer cylindrical masticator is jacketed and heated with steam:

COMPOUND NO. 1.

Pure cleaned hard Gutta-percha	28
Pure cleaned tough selected Gutta-percha or Balata (preferably more rather than less)	11
Pure cleaned "low white" Gutta-percha (preferably less rather than more)	9
"Crumb" or ground good old vulcanized India-rubber	34
Hardwood veneer dust	5
Sulphur	6½
Zinc oxide (or zinc dust)	3¼
Flocking, or the cut fiber of cotton textile fabrics	3¼
Total	100

COMPOUND NO. 2.

Pure cleaned tough Gutta-percha	8½
Pure cleaned Balata or selected Gutta-percha	8½
Pure cleaned "low white" Gutta-percha	24
"Crumb" or ground good old vulcanized India-rubber	33
Hard ground veneer dust	5
French chalk, powdered	6
Sulphur	6
Zinc oxide (or zinc dust)	3
Flocking, or the cut fiber of cotton textile fabrics	3
Alum, ground	3
Total	100

Another compound patented by Mr. Dick embraced the use of low grade African and Borneo rubbers, which, after cleansing, were mixed with Gutta-percha while still moist in hot water. After the mixing the compound is treated under a moist heat, where the temperature is 212° to 240° F., the result being a tough, plastic, fibrous dough. This compound is then, so the inventor claims, equal to any service for which the Gutta-percha and Balata compounds are used. An important property in this compound is the shrinking quality which Gutta-percha possesses, while its power of cohesion rendered it especially valuable for insulating wires.

Shepard mixed Gutta-percha with sulphur, exposed it to a heat varying from 300° to 350° F., admitting hot air, then combined it with sulphur and earthy matters. It was then vulcanized by Parkes's cold curing process.

Parkes dissolved Balata and mixed it with 5 per cent. of chloride of sulphur, diluted with mineral naphtha. Gun cotton was also dissolved to a pasty mass, in naphtha distilled with chloride of calcium, and the two solutions were combined, forming a soft, flexible compound.

Childs vulcanized Gutta percha by mixing it with sulphur and placing it in a vulcanizer containing hydrated lime, and then turning on heat sufficient to obtain enough steam from the lime to do the curing.

Duvivier and Chaudet treated Gutta-percha with bromide of sulphur or chloride of sulphur, making it more elastic and less liable to be acted on by heat or cold. When acid vapors were formed during the operation, carbonate of sodium was mixed with the solution.

Rostaing made Gutta-percha hard and unalterable by treating it, after cleansing, with caustic soda, which was thoroughly washed out. after which it was combined with silicate of magnesia and treated with tannin, catechu, and other astringent matter.

Keene cured Gutta-percha articles by exposing them to the fumes of sulphur or immersing them in a bath of melted sulphur.

Charles Hancock treated Gutta-percha in a bath of boiling water in which was carbonate of potash, or muriate of lime, leaving it for an hour, and then mixing it with lead, glue, and bitumen. His claim was that this treatment hardened the Gutta-percha, rendered it better adapted for bearing friction, and less likely to be oxidized. He also cured Gutta-percha by mixing with it sulphur, sulphides or orpiment, and applying heat. He gave as a compound for vulcanizing Gutta-percha 48 parts Gutta-percha, 6 parts golden sulphuret antimony, and 1 part sulphur, the compound to be boiled under pressure.

Emory Rider mixed Gutta-percha with oxide of lead, heated

it in open steam heat until the oily matters were expelled, then mixed it with hyposulphite of lead and cured it.

Lucas prepared a printing roll of Gutta-percha, first immersing the Gutta-percha in nitric acid, and then placing it for an hour in a solution of carbonate of soda, thus producing a tougher wearing surface.

Barlow and Forster mixed Gutta-percha with Kauri gum and milk of sulphur for a cable coating.

Macintosh immersed Gutta-percha in concentrated sulphuric acid for a number of seconds to harden the surface. He also mixed Gutta-percha with gun cotton, curing with sulphuric acid, claiming that the resultant compound was not likely to be affected by the heat of tropical climates.

Analyses of common Gutta-percha, by Edouard Heckel and Fr. Schlagdenhauffen:

Gutta	75	to	82
Albane	19	to	14
Fluavile	6	to	4
Total	100		100

Analysis by Payen:

Gutta	78	to	82
Albane	16	to	14
Fluavile	6	to	4
Total	100		100

Gutta-percha is made of a mixture of hydrocarbons, and there is usually present a certain amount of oxygen. According to Granville H. Sharpe, F.C.S., its ultimate composition is:

Carbon	86.36
Hydrogen	12.15
Oxygen	1.49
Total	100.

[Specific gravity, 0.96285 to 0.99923.]

The primary analysis of Gutta-percha by Sharpe is:

Hydrocarbon	79.70
Resin	15.10
Wood fiber	2.18
Water	2.50
Ash	0.52
Total	100.

Obach gives the following average results from a large number of analyses of each of twelve leading brands or sorts of Gutta-percha:

	Gutta.	Resin.	Dirt.	Water.
Pahang	78.1	19.2	1.5	1.2
Banjer red	67.0	30.2	1.5	1.3
Bulongan red	68.6	29.0	1.4	1.0
Bagan	57.5	40.9	1.0	0.6
Goolie red soondie	55.2	42.9	1.2	0.7
Serapong	56.2	42.4	0.9	0.2
Bulongan white	52.2	45.4	1.5	0.9
Mixed white	49.8	47.4	1.1	1.7
Banjer white	51.8	44.1	1.8	2.3
Sarawak mixed	55.6	40.9	1.8	1,7
Padang reboiled	50.3	45.8	2.0	1.9
Banca reboiled	46.8	51.1	1.1	1.0

Another series of analyses by Obach relates to the constitution of the resins in Gutta-percha, as follows:

	Albane.	Fluavile.
Carbon	78.76	80.79
Hydrogen	10.58	11.00
Oxygen	10.46	8.21
Total	100.	100.

Some typical Gutta-percha cement compounds follow:

1.—For joining wood: Gutta-percha, 11 pounds; shellac, 3 pounds; Venice turpentine, 5 pounds; pitch, 1 pound.

2.—For uniting metals, glass, stone, and earthenware: Gutta-percha, 45 pounds; shellac, 20 pounds; gum mastic, 5 pounds; oxide of lead, ½ pound; storax, 3 pounds; Venice turpentine, 26½ pounds.

3.—For cementing leather: Gutta-percha, 4 ounces; bisulphide of carbon, 20 ounces; asphaltum, 1 ounce; common resin, 1 ounce.

4.—Gutta-percha glue: Gutta-percha, 1 pound; rosin, 1 pound; litharge, 1 ounce; powdered glass, *quantum sufficit.*

5.—Shoemaker's wax: Melt Gutta-percha, 20 ounces; add pitch, 58 ounces; soap, 5 ounces; rosin, 6 ounces; beeswax, 5 ounces; palm oil, 1 ounce; tallow, 5 ounces.

6.—For preserving metals and other surfaces: Coal tar, 20 pounds; Gutta-percha, 5 pounds; minium, 6 pounds; white lead, 7 pounds; pitch, 10 pounds; resin, 10 pounds; spirit turpentine, 4 pounds; sulphur, 38 pounds.

7.—General cement: Make a solution of Balata of 5 ounces in ¼ gallon naphtha, and another of Gutta-percha 5 ounces in ¼ gallon naphtha. Combine the two solutions and add 13 ounces resin or pitch and stir and mix thoroughly.

THE ANALYSIS OF GUTTA-PERCHA.

THIS of course refers to the analysis for the crude gum, and, to have the analysis complete, it should cover the amount of water present, the amount of foreign matters and impurities, the amount of ash, the amount of pure gutta, and the amount of resins.

The water is easily determined by heating a known weight from the sample at a temperature ranging from 212° to 230° F., the loss in weight being the amount of water present. This is a common process in chemical analysis. In the case of Gutta-percha, it must be varied, as the sample is liable to oxidize even under examination causing an increase of weight. This is overcome by conducting the heating in a slow current of nitrogen, or carbonic acid gas.

J. A. Montpellier devised an apparatus for this, which consisted of a special retort with a large opening which he used as a vapor bath and having a tubulure at its side. It is closed by a large cork, in which there are two holes, one for the tube which is to introduce the gas, and the other for the thermometer. The sample to be dried is placed in a crucible of porcelain or platinum suspended within the retort. As the water evaporates it is borne by the current of gas through a tube inserted in the side tubulure, and into U-shaped tubes, containing sulphuric pumice, which retain it. Further on the U tubes are connected with a Liebig tube with five bulbs containing pure sulphuric acid preventing the entrance of moist air after the apparatus cools, a further use being to make it possible to regulate the speed of the current of gas.

The retort is immersed in an oil bath heated by a Bunsen burner. If carbonic acid is used it is obtained by the action of hydrochloric acid on marble chips produced in a Kipp apparatus followed by wash flasks, the first of which contains bicarbonate of potassium in solution, which is intended to stop the passage of any hydrochloric acid, and the second containing sulphuric acid at 150° to thoroughly dry the gas. To be absolutely sure that this gas is dry, a dessicator filled with sulphuric pumice is placed between the retort and the second wash flask. The operation of drying one gram with this apparatus, takes 6 or 7 hours. The determination of the amount of impurities which comes next may be effected very easily, by using M. F. Jean's exhaust apparatus. A small part of the sample, from one-half a gram to a gram, is weighed, cut into small fragments, put in a filter, the weight of which is known, which in turn is placed in a platinum cone. This cone is then put in the extension of the apparatus; this extension communicates by two tubes with the retort containing pure chloroform. A condenser, in which a current of cold water constantly circulates in order to condense the chloroform vapor, is placed at the upper part of the extension.

The retort rests on a sand-bath, very gently heated by a Bunsen burner. Under the influence of the slight heat the chloroform evaporates, passes through one of the tubes, and drops on the filter containing the Gutta-percha, which it gradually dissolves. The solution, passing through the filter, then drips into the retort through the second tube.

All the impurities remaining in the filter, it is sufficient to dry and weigh the filter to get the weight of the foreign matters, the drying should be done in the apparatus used in determining the amount of water.

The next process is the determination of the amount of ash. In Gutta-percha this is always very small, as mineral matter is almost entirely absent from it, the quantity never exceeding one-half of 1 per cent. The amount of ash is determined by burning in a capsule of platinum or porcelain a known weight of Gutta-percha.

The fourth step is the determination of the amount of pure gutta, and of the resins. Both fluavile and alban are soluble in absolute alcohol at the boiling point, and as pure gutta is insoluble in it, this is a very ready means of separation. The sample to be examined is cut in little bits, put in a platinum basket which is pierced with holes, and hung in a retort containing the alcohol. This retort is heated with a sand-bath or water bath, the vapor of the alcohol passing through a Liebig condenser and returning to the retort. The boiling is continued for 5 or 6 hours, with the basket immersed in the alcohol. It is then raised above the liquid, and the boiling continued for 5 or 6 hours more. The latter part of the process removes the last traces of resin.

The boiling operation being completed, the pure gutta together with the impurities remains on the filter. There remains then the drying of the filter in the apparatus used in determining the amount of water and the weighing of it. The loss of weight shown by the Gutta-percha corresponds to the amount of resins increased by the weight of the water. Subtracting that weight, which has already been determined, the weight of the resins remains.

INDEX.